IEE Control Engineering Series 18

Series Editors: Prof. B.H. Swanick
                Prof. H. Nicholson

APPLIED CONTROL THEORY

Previous volumes in this series:

APPLIED CONTROL THEORY

J.R.Leigh

PETER PEREGRINUS LTD
on behalf of the
Institution of Electrical Engineers

Published by: The Institution of Electrical Engineers, London
and New York
Peter Peregrinus Ltd., Stevenage, UK, and New York
© 1982: Peter Peregrinus Ltd.

British Library Cataloguing in Publication Data

Leigh, J.R.
   Applied control theory. — (IEE control
   engineering series; 18)
   1. Control theory
   I. Title   II. Series
   629.8'312   QA402.3

ISBN 0-906048-72-9

Printed in England by Short Run Press Ltd., Exeter

Contents

Preface

Every application of control theory to a real situation poses certain basic problems. What is the overall objective? How does the theory interface with economic, human, software and measurement factors — compromise and modification will be needed to take these into account.

In such a situation, no-one would hope to suggest a structured applications methodology. However, based on experience, I have developed an approach that is summarised below and which governs the structure of the book.

(a) Economic aspects must be considered at an early stage of any project.
(b) Simple techniques and ready-made manufacturers' solutions should be applied wherever possible.
(c) More advanced techniques will be received enthusiastically in those applications where they can offer a genuine contribution.
(d) Control systems using distributed microprocessor power will have an impact that is difficult to exaggerate. Control engineers must become familiar with the concepts involved.
(e) Familiarity with a range of applications is indispensable in developing an efficient approach in the field of applied control theory.

Within this structure, I have described techniques that are useful or potentially useful. To satisfy my remit, I have often stated the obvious but, less frequently, have included advanced concepts where necessary.

The book should be accessible to a wide variety of engineers. Preferably they should have an elementary knowledge of automatic control theory.

Acknowledgements

The material that I present in this volume represents the fruits of many co-operative projects. I acknowledge my debt to a succession of co-workers with whom it has been a pleasure to be associated.

A number of people have given permission for their work to be described. In particular, I should like to thank the following:

Dr. H. J. Wick of the Automation Engineering Department, Estel Huttenwerke Dortmund AG, West Germany, for permission to quote the results of his application of the Kalman filter (Section 3.6).

Dr. S. A. Billings of Sheffield University, for permission to base the heuristic explanation of the Kalman filter on a previous presentation of his.

Dr. K. W. Goff of the Leeds and Northrup Company, U.S.A., for supplying and allowing me to use the graph, Fig. 5.6.

Dr. O. L. R. Jacobs of the Department of Engineering Science, University of Oxford, for supplying and allowing me to quote results comparing the performance of different types of algorithms (Section 5.3).

The Foxboro' Company, Redhill, England, for supplying an abundance of application oriented literature with permission to quote from it.

I thank the companies whose approaches to process control using distributed computers are discussed in Chapter 9. They readily supplied information and invited me to see their equipment in use.

I thank my colleagues, Mr. Tony Thornton, for drawing Fig. 10.16 and Mr. Ken Pallant for reading carefully through the manuscript.

Finally, I wish to thank Mrs. Geraldine Coveney for typing the manuscript and my son, Ian, for designing the cover.

Chapter 1

Introductory topics

1.1 Understanding the process

In this book, the 'system to be controlled' will normally be referred to as the *process*. (The choice of word should not be taken to imply a restriction of the interests of the book to the traditional process control area. Alternative words such as mechanism, would be equally valid.)

To the surprise of some undergraduates, processes do not carry labels marking their variables nor, alas, are they conveniently pre-classified into linear, nonlinear, stochastic, etc., etc., categories. The ability to come to terms with this situation is a pre-requisite for anyone proposing to succeed in an industrial environment.

Consider the two diagrams, Figs. 1.1 and 1.2. The first is the standard text-book representation of a controlled process; the second is more realistic, since it allows for product in and out flow, measurement, actuators and operator–machine communication.

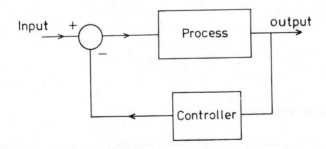

Fig. 1.1   *A conventional text book representation of a control loop*

A *continuous process* is one that, for a significantly long period, receives a nominally constant input and is required to produce a continuous constant output. The control problem is to maintain the product in specification despite the effects of disturbances.

A *quasi-continuous process* is a process that, in form, is identical to a continuous process but that runs only for short periods. The reasons for the operation in short time periods may be imposed by the process itself or they may be implemented by management for reasons of convenience. The control problem has three parts:

(*a*) To start up the operation quickly and efficiently.
(*b*) To control during the operation as for a normal continuous process.
(*c*) To finish the operation efficiently.

A *batch process* is one where, typically, a set of constituents is mixed together in a container at time zero and then processed until it has been converted into a desired product. The container is emptied and the procedure is repeated. Batch processes and their control are discussed in Section 10.14.

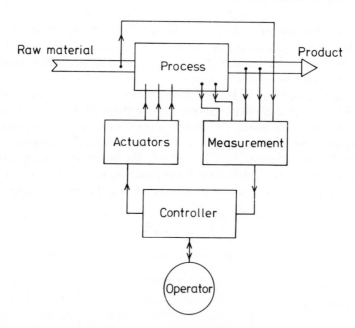

Fig. 1.2   *A more realistic representation of a control loop*

In terms of equations, a batch process is described by a set of *n* ordinary differential equations.

$$\dot{x}_i = f_i(x_1, \ldots, x_n, u_1, \ldots, u_r), \quad i = 1, \ldots, n.$$

Initial conditions $x_i(0)$, $i = 1, \ldots, n$, and desired final conditions, $v_i, i = 1, \ldots, n$, are given. The control problem is to steer the process so that for some time T, the equations $x_i(T) = v_i$, $i = 1, \ldots, n$, are satisfied as nearly as possible. The strategy over the period from 0 to time T must be such that overall costs are minimised; in particular, T itself enters the calculation for minimum cost.

A continuous process, by contrast, is described by a set of partial differential equations of the form:

$$\frac{\partial x_i}{\partial t} = f_i(x_1, \ldots, x_n, u_1, \ldots, u_r), \quad i = 1, \ldots, n,$$

$$\frac{\partial x_i}{\partial l} = g_i(x_1, \ldots, x_n, u_1, \ldots, u_r), \quad i = 1, \ldots, n,$$

where l ranges over the values 0 to L, where L is the length of the process. Initial conditions, $x_i(0, l)$, $i = 1, \ldots, n$, at time 0 and for all values of l along the process, are given. Desired values v_i, $i = 1, \ldots, n$ are given as before for the batch process. The control problem is to maintain the process end values equal to the desired values, i.e. to satisfy the equation

$$x_i(t, L) = v_i, \quad \text{for all values of } t.$$

1.1.1 Comparison of continuous and batch processes

Quite apart from control considerations, nearly every process benefits economically from conversion from batch to continuous operation, provided that demand is sufficient to keep the continuous process in operation.

Control of a continuous process is usually simpler than control of a batch process. The reason is that the main control objective is to maintain the outputs constant at all times. Deviations can be corrected by simple controllers based on linear assumptions. Batch processes have a number of special control problems. They have to be steered from an initial condition to a desired final condition in the most efficient manner. Often this means following a pre-determined strategy and the role of feedback control is to ensure that the strategy is followed.

1.2 The effects of process characteristics on attainable performance

How well can a particular process be controlled? How quickly can it be brought from one state to another? How well can it track a particular given desired value curve? How good a steady-state accuracy can be secured? It is important to realise that the process itself sets the limits on what can be achieved and the mystique surrounding controller design should not be allowed to obscure that fact.

Starting with the simplest possible case, let the process be a capacitor in series with a resistor. The shortest time in which the capacitor can be brought from zero charge to a particular state of charge is determined by the values of resistance and capacitance in conjunction with the maximum applicable voltage – all three being part of the process specification. All that a controller can do is to implement a strategy to achieve this shortest time.

Thus it is that the process time constants taken together with the knowledge of constraints on the input variables determine the maximum response rate. In fact the actual response rate achieved will usually be a great deal less than the theoretical maximum. Process dead-time will restrict the usable closed-loop gain often to a very low value. Steady-state error depends on the zero frequency gain in the control loop. A limit to the closed-loop gain will always be set by either stability or noise considerations.

A knowledge of process order, time constants and dead-times, together with operating constraints, allows, with practice, estimates to be made of achievable performance. If these performances are not acceptable, then the process must be modified by interaction between control personnel and process designers or managers. Many examples can be quoted from actual situations:

(*a*) The inertia of a shaft was reduced to improve the response of an electrically driven actuator.
(*b*) A new by-pass oil pipe was laid specifically to allow satisfactory oil viscosity control.
(*c*) Rolling-mill mechanical tensioning arms were re-designed to have reduced inertia to achieve faster response.
(*d*) Power electronic circuits were re-designed to improve the response of large DC motors.

Some processes possess a considerable degree of self-regulation even in the absence of control. If subjected to perturbations, they tend to settle naturally at a new operating point. Such behaviour is a function both of dynamic structure and of favourable nonlinearities.

Thus the temperature of an electrically-heated oven tends to be self-regulating. If the applied voltage increases, the temperature will rise to a level where losses balance the increased energy input. Because of the nonlinearity of the heat loss *versus* temperature curve, the temperature will rise less than forecast by a linear calculation. Self-regulating processes can be expected to be rather simple to control.

Processes that are open-loop unstable generally pose a demanding control problem. Broadly speaking, the control exercised must be powerful enough and fast enough in action to overcome the inherent instability in the process. A good example is an inverted pendulum – it can only be stabilised by fast acting powerful control actions.

1.3 The interaction of control with process design and operation

Most closed control loops are necessary to minimise the undesirable effects of some disturbing influence. In process control, the disturbances either enter with the raw material (in which case they can be considered as due to some shortcoming in the control of the previous process) or they arise within the process itself. Now it is generally true that most of the disturbances can be compensated for either by improvement in plant design or operating practice, or by implementing a suitable control scheme.

Why state this? Simply that one hundred percent reliance on control to remove the effects of disturbances will almost certainly fail, as will one hundred percent reliance on plant design modifications. A good minimum-cost working solution to reduce the effect of disturbances will usually have some elements of process modifications as well as an element of control. This implies that control engineers must be involved in the formative stages of plant design and that they must be fully aware of process scheduling and operation if they are to make a useful contribution.

1.4 The role of the operator in control schemes

Few, if any, industrial control schemes operate without supervision, interaction or intervention by human operators. It follows that such interaction should be designed into the system with proper regard for the skills of the operators.

Human operators are at their strongest, in comparison with automatic controllers, in unexpected situations for which there is no exact precedent, in dealing with information that is difficult to quantify or for which no sensor exists, and in general in situations demanding adaptivity, pattern recognition and 'feel'. Automatic loops are superior where very fast actions are required, where very slow actions are required, and where the actions are well-defined or repetitive. In order to make best use of the matchless (and expensive) skills of an operator he must be presented with first class, well-displayed information.

As an initial reaction it is tempting to consider simulating a proposed control system complete with its human operators. In this way the tasks delegated to the operator could, in theory at least, be modified until best performance was obtained. The problem with this approach is that human operators are unexpectedly complex, even when dealing with trivial tasks.

For instance, let the very simple case be considered where the operator takes in visually a single signal and moves a single dimensional control lever. Even here, three parallel models (pursuit, pre-cognitive, compensatory) are needed to represent perception and the relative contribution of these models depends on circumstances. The human response model needs to be completed by another complex block representing the neuromuscular activation system — itself a feedback system.

Interesting experiments, McRuer (1980), have shown that if a human operator is required to control a process of transfer function

$$G_1(s) = \frac{k_1}{s(1 + sT)}$$

and is given time to learn how to do this, then after the learning period his transfer function can be measured to be

$$G_2(s) = k_2(1 + sT)e^{-s\tau},$$

i.e. the operator's transfer function contains a numerator term compensating the first-order lag in the process. The time τ is not a constant but depends on the task demanded of the operator. In particular, the more the operator is required to act as a lead network, the greater the time-delay τ becomes. A typical figure quoted for τ by McRuer is 0·16 seconds.

Thus, the very adaptivity of the human operator prevents his being neatly represented by a transfer function except in trivially simple cases that are not of real interest.

Chapter 2

The economics of industrial control

2.0 Introduction

The first stages of any large process control project should almost always include an economic assessment. The first aim of the assessment is to enable firm cost benefit forecasts to be laid before management to persuade them to commit expenditure to the project. The second, less tangible but equally important aim, is to allow the project team to obtain a feel for the relative importance of the factors involved.

The techniques for economic assessment that are described below have each been found useful in particular industrial situations. Attention is concentrated on quantifying the benefits of control. The cost of implementing a control scheme can be estimated by obvious methods, so this aspect is not considered here. Only one point needs to be made. The low cost of microprocessor boards does not, in general, mean that low-cost process control can be achieved, since sensors, cabling, displays, software and engineering effort all need to be taken into account.

The performance of a process depends on its design, on operating practices and on the level of control that can be achieved. The three areas interact strongly in their effect on performance. An economic assessment of a process will often show that improvements are called for in all three areas (process design, operating practice, control system) if maximum benefit is to be obtained from a particular level of investment.

Papers describing techniques for economic assessment are relatively rare. See Funk and Smith (1974), H. W. Smith (1977), Stout (1973) and Gordon and Spencer (1977).

2.1 The economic profile of a process

It is impossible to work realistically on the control of a process without having some order of magnitude feeling for the relative economic importance of the main process variables. To help to obtain this knowledge, a simple economic model of the process can be used as shown in Fig. 2.1. The percentages indicate how the value of

x pounds/ton is made up. For instance, 70% of the value is contributed by raw material. (The value of the product is, of course, its internal value and not its market value.) While it is not possible to use this model quantitatively it gives an economic sense of proportion to a control project.

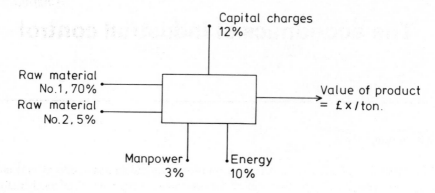

Fig. 2.1   *A simple economic model of a process*

Simple models like that of Fig. 2.1 can be connected together to represent a complete plant in which all the inputs are referred to the final output. This allows the relative importance of the main process variables to be appreciated in the context of the complete plant.

There is one way in which models like that of Fig. 2.1 can mislead if insufficient thought is given to their interpretation. Consider the particular numbers given in the figure. It may appear that a control system that attempts to reduce the use of raw material 1 would have considerable potential. However, the composition of the output product might decree that, inevitably, the main cost is bound to be raw material 1. Perhaps the biggest possible savings might be in energy. It is therefore worthwhile to consider, for each variable, whether there are irreducible minima (as in the case of the raw material − implied by the law of conservation of mass). For heating processes the irreducible minimum energy would then be that to satisfy the heat content in the heated product − all other energy being regarded as operable on by a control system to produce savings.

2.2 The economic justification of control schemes

Industry will only invest effort and expenditure in a project when there is a sound economic justification that, so far as possible, supports the decision to invest. Not every project can be guaranteed to be a success and this factor must be borne in mind, particularly when considering high risk projects. In general, the payback on investment if the project is successful, multiplied by the probability of success, must be high if investment is to be justified.

Economic justification for control systems tend to be a weighted combination of various roughly quantifiable factors, together with descriptive factors that go forward unquantified. The table below gives typical factors for process plant.

Quantifiable factors	Descriptive factors
Improved product uniformity	Improved market competitiveness
Increased yield	Improved safety
i.e. $\left(\dfrac{\text{mean output rate}}{\text{mean input rate}}\right)$	Reduced emission of noise or atmospheric pollution
Increased throughput	
Reduced energy consumption	
Reduced manning requirement	

2.2.1 *Estimation of cost benefits*

Stout (1973) lists the following ways in which the cost benefits of process control can be estimated for a future installation.

(*a*) Experience from a similar installation.

(*b*) (For an existing process that is under manual control.) Experience of the best that can be achieved under manual control.

(*c*) Theoretical calculation of the best possible performance. This is a 'negative criterion' that can be used to reject a proposed automation scheme if the best possible performance is too low.

(*d*) Simulation of the proposed scheme. Such simulation requires a mathematical model of the process and of the control scheme. Simple models will often suffice.

(*e*) Manual trials of the strategy. These trials, which can augment economic assessment by more theoretical methods, can usually be conducted provided that (i) the necessary sensors and displays are fitted to the process, and (ii) the actions required are not too rapid or too complex for an operator to follow.

(*f*) A full-scale temporary trial. For a large scheme the cost of a full-scale trial will be high and this option should only be adopted after careful costing.

(*g*) Evaluation at a pilot scale plant. Pilot plants do not, in general, provide completely satisfactory testing conditions for control schemes to be used on full-scale plants. This is because many engineering difficulties and problems caused by operating practices cannot be represented realistically.

2.3 Attaining control level zero

Many industries still have items of plant operating with little or no control. They may be operating perfectly in this way, in which case nothing further needs to be said, or they may be capable of significant improvement by the application of control. The most common situation is that process variables are fluctuating widely but that a financial benefit cannot be put on the reduction of variation. Reducing the variation of process variables has been called control level 0 by some workers. By itself, control level 0 may have little or no financial benefit (even though it may be costly to attain), but it is the pre-requisite for implementing 'higher level' control systems that do have financial benefit. As a simple example (illustrated in Figs. 2.2 and 2.3), consider the manufacture of a consumer product that is required by law to weigh at least 454 g. To allow for variation, the manufacturer finds it necessary to dispense the product at a mean weight of 460 g. Level 0 control reduces the variability but brings no immediate benefit. However, higher level control moves the mean weight to 455 g and produces cost benefit.

In this instance, once level 0 control had been attained, higher level control simply consisted of moving the set point of the mean weight.

It is generally true that the initial investment in instrumentation and control will be spent in bringing the plant to stage 0 — not in itself profitable — but a necessary pre-requisite for higher levels of control.

2.4 The desirability of adding a new or improved control scheme to an existing plant

A surprisingly large proportion of the time of an industrial control engineer is spent in this area of work, since it is still rare for items of plant to be delivered complete with control as an integral part, such that no improvement needs to be sought. For a plant where the new control scheme aims to reduce the proportion of unsaleable product, a very direct method of economic justification exists. This is illustrated in the two histograms, Figs. 2.4 and 2.5. Figure 2.4 gives the present performance of the plant while Fig. 2.5 gives the expected performance, with the proposed control scheme. The shaded area can, relatively easily, be converted into financial savings and these, together with the cost of the control system, can form the basis for an economic assessment. Figure 2.4 has to be produced by mounting a carefully controlled discard measuring trial over a sufficiently long time period. Existing accounts data will rarely suffice since they tend to relate to the overall performance of a whole plant. The second histogram, Fig. 2.5, needs to be produced by simulation. How complex does the mathematical model need to be for this simulation and what inputs should be assumed? There are no firm rules and the following guidelines are based on experience. The model should include only first-order effects. Transients can often be neglected, so that the model may reduce to a set of algebraic equations with constraints. Inputs should ideally be members of the identical set from which the first histogram was produced. Thus, histogram 2.5 is produced in the pedestrian fashion of repeating, by simulation, the trials previously measured in practice, but this time assuming that the proposed control scheme is operational. If three or four possible control systems are under consideration, this can mean a lengthy computational task. It may be worth considering an alternative approach in which

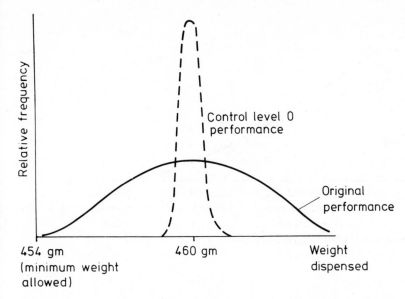

Fig. 2.2   *Improved control reduces the deviation of the product from specification*

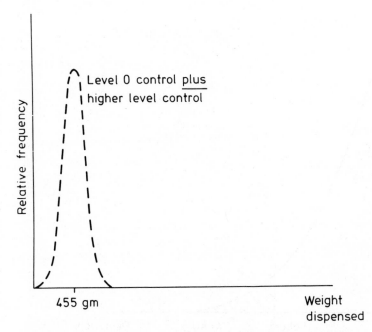

Fig. 2.3   *Improved control plus movement of the mean increases profitability*

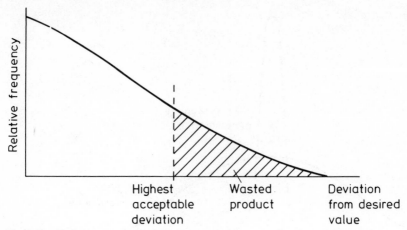

Fig. 2.4   *Performance of existing control system*

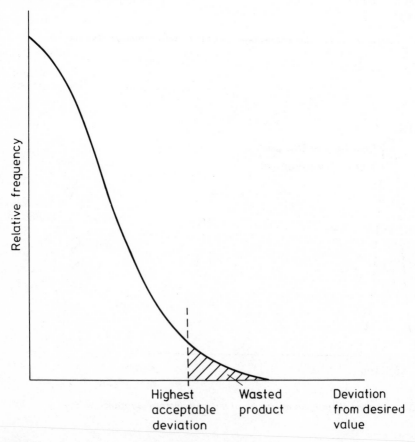

Fig. 2.5   *Performance of proposed control system*

parameters characterising the input statistics are input to a statistical model of the process and its proposed control scheme. The output results are then parameters describing the shape of the required histogram. The author has experimented with this approach using input and output power spectral densities of error, which can be related through the transfer function of the plant with its proposed control system. In practice, many idealising assumptions usually have to be made before the approach can be applied, so that in most investigations the earlier, slower method will still have to be used.

2.5 The use of curves of return against investment

If a control engineer proposes a technically viable control scheme and supports it by a sound economic justification that predicts a good return on investment, surely he is performing his duties perfectly? Not necessarily, for the scheme may be unnecessarily sophisticated. This is illustrated by referring to a hypothetical case where a simple criterion of suitability for investment is being used: that the annual return on capital invested should be greater than or equal to $x\%$. From Fig. 2.6 it is clear that the proposed scheme gives a more than sufficient return on investment. Now take a more detailed examination to produce Fig. 2.7, where the same scheme is broken down into two phases, 1 (crude), and 2 (sophisticated). It is at once apparent that phase 2 is an unjustified luxury. The plant manager should be advised to invest only in phase 1.

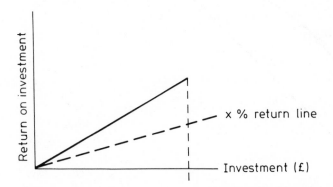

Fig. 2.6   *Return on investment — no subdivision of the investment*

The example illustrates the usefulness of curves relating return with investment.

A return *versus* investment curve such as the one in Fig. 2.8 can be sketched out for a particular automation proposal, once simulations, as described in Sections 2.2 and 2.4, have been completed.

On the curve, two points A and B have been marked. A is the point of maximum slope. B is the point of unity slope. Clearly, point A yields the best rate of return

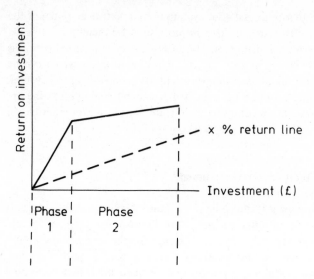

Fig. 2.7   *Return on investment — with subdivision of the investment*

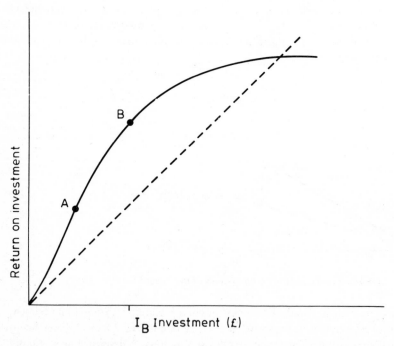

Fig. 2.8   *Return on investment — points* A *and* B *indicate maximum rate of return and maximum justifiable investment respectively.*

on investment. Any investment greater than I_B cannot be justified since the increment of expenditure greater than I_B will not yield even a unity return.

2.6 The time factor in the economics of control schemes

Assume that a new automation scheme will cost £X and that the scheme, when implemented, is expected to make a profit of £$X/3$ per year. It seems that the scheme will 'pay for itself' in three years — a good investment? Now inject realism into the assumptions: the £X to pay for the automation scheme has to be borrowed at a rate of 15% p.a. The expected improvement in performance is not achieved until $1\frac{1}{2}$ years have elapsed after investing the sum of £X.

The following table shows the simplified balance sheet of the operation.

In the table, an initial investment value of $X = 100$ has been substituted.

Years elapsed	Credit	Debit	Overall balance
0	0	100	−100
1	0	15	−115
2	17	17	−115
3	33	17	− 99
4	33	15	− 81
5	33	12	− 60
6	33	9	− 36
7	33	5	− 8
8	33	1	+ 24

It can be seen, using the more realistic assumptions, that the scheme needs more than seven years to pay for itself.

Two lessons are to be learned from this simple example. Bank interest rates have to be considered and the time taken for the scheme to begin paying back must be allowed for. This shows clearly that failure to meet implementation time targets can upset the validity of a scheme.

2.6.1 Putting a cost on time

An economic case for new investment once stated, in summary, 'This scheme will speed up the process by 5%'. Current production is x tonnes/year, new production will be $1 \cdot 05x$ tonnes/year. The case then went on to turn the $0 \cdot 05x$ tonnes/year into cash. What is wrong with this reasoning? Simply that it assumes the sale of 5% more product (or, if the case applies to an individual item of plant embedded in a factory, that the item of plant is at present producing too little). Without spelling out all the possibilities, it is clear that, if the plant is not fully utilised, then the cost benefit of a scheme to increase throughput rate will be much less than might have been calculated on such a naive assumption. Only in special cases will it be true that an increased available throughput rate will be used by management to increase total throughput. Thinking can be clarified by assuming that the plant is not yet built. In this case, for a particular required annual output, management can decide to build a plant of a particular size without the control scheme or to build a somewhat smaller plant with the control scheme. It is now apparent that savings in time are related to capital cost. Such reasoning leads to the point where we argue that a control scheme to increase throughput rate is justifiable only if it is cheaper than a scheme simply to increase the size of the plant. Cases will always have to be considered individually. Existing plants often cannot be enlarged and time scales of implementation are vitally important. Nevertheless, the viewpoint that is advocated in this section is a healthy initial stance. Indeed, a NATO guide-line paper on process economics defines a general performance index for process plant of the form:

$$I = \text{throughput rate}$$

$$\times \left(\frac{\text{value per tonne of output product} - \text{cost per tonne of input}}{\text{capital charges}} \right).$$

Here it is seen that throughput rate and capital charges appear as a ratio — supporting the argument.

2.6.2 *Allowing for the time taken for a new control system to achieve its design performance*

Experience shows that a large complex control scheme takes time to achieve the level of performance envisaged by its designers, even in the absence of implementation problems. A curve like that shown in Fig. 2.9 is typical for an existing plant whose performance is to be improved by the addition of a complex scheme. The plant is supposed to operate at performance level A originally and at level B with the control system working fully. The time scale will depend on the complexity of the scheme. It is clear that an economic justification based on an instantaneous step jump in performance is unlikely to be realised in practice. Clearly, monetary inflation makes matters worse and has to be taken into account.

2.7 Economic choice of the best configuration of a sequence of controlled plants

Consider the frequently occurring situation where raw materials enter a factory and pass through several processing stages before being converted into a final product.

The apparently simple question arises: how should investment in control be optimally apportioned between the different stages of the sequential process?

For simplicity, assume that the quality of control at any stage j can be represented by a single parameter q_j. Designate the estimated costs of n different possible control schemes at stage j by c_{jk}, $k = 1, \ldots, n$. If there are m process stages then the total cost of the control for a particular configuration is

$$C = \sum_{j=1}^{m} c_{jk},$$ the correct value of k being substituted to correspond with each value of j.

For purposes of this exercise, we argue that q_j depends on q_{j-1} and c_{jk}

$$q_j = f_j(q_{j-1}, c_{jk}),$$

(i.e. the quality of control achieved at the jth stage depends on the quality of the incoming product and on the size of the investment in control).

Thus the quality of control achieved at the last stage is

$$q_n = f_n\{f_{n-1}[\cdots (f_1(q_0, c_{1k})) \ldots]\}.$$

If approximate process models exist then the functions f_j can be simulated numerically.

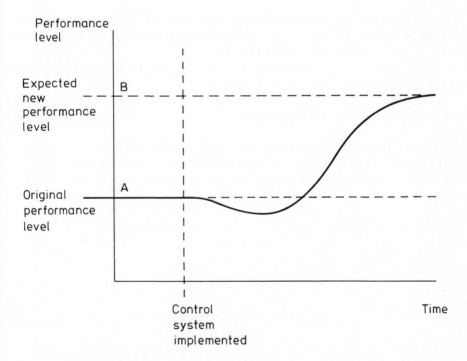

Fig. 2.9   *Improvements are not achieved instantaneously but typically follow a curve of the form shown*

For a particular desired value of control q_n in the final product, there will in general be many particular choices of control configuration. However, we can search for the minimum cost value of C that will achieve such a q_n. Basically, this requires a substitution of all possible values of k for each value of j. Readers will recognize the influence of dynamic programming in the nested equation for q_n and the computation can be achieved efficiently using a dynamic programming technique. At the end of the exercise, management can be given a set of possible product qualities $q'_n, q''_n, q'''_n, \ldots$, and for each, the optimum configuration for its realisation and the total cost can be quoted. Thus the technique produces not one, but a whole series of recommendations. Each recommendation is an alternative that management should consider in relation to market requirements and availability of investment funds. This is an optimisation technique that operates in the real world!

Of course, most sequential process routes are not so simple as in the example. The process route will often fork or even have closed loops but with care the approach advocated here will still be useful in deciding how much to invest in control and how to balance the investment between the different stages in the sequence.

2.8 Post-implementation checks on economic benefits

Checks to ensure that the hoped-for benefits of an automation scheme are, in fact, obtained may be carried out by techniques similar to those described earlier as suitable for the pre-implementation justification.

Experience shows that short trials of prototype control schemes sometimes give over-optimistic results that cannot be maintained on a routine basis. (For instance, a system that is not robust enough to work consistently can be made to give good short-term results by special efforts such as may be expended during a prototype trial.)

During their lifetimes most control schemes undergo a slow but significant evolutionary improvement. Such improvements can be encouraged by having a flexible, extendable control scheme that can be understood by process operators and engineers.

Chapter 3

Measurement for control

3.0 Introduction

Good measurement is a vital pre-requisite for worthwhile control to be achieved. Lack of suitable primary measurements has, from the outset, limited the scope of process control systems.

Measurement of primary variables in large-scale industry is very costly and any investment made will be contingent on an expected payback. Economic calculations can only be made in an overall systems context since it will be in improved system performance that any payback will materialise.

Conventional measurement topics are not covered here but a number of comprehensive handbooks and specialist monographs are listed in the bibliography. Attention has instead been concentrated on a number of important topics that fall in the grey area between measurement and control.

One of the most important tasks in control system design is to select the variables that are to be measured and controlled. Economically important phenomena are often not representable by simple physical variables. Examples are the strength of paper or the flatness of a steel plate. Both properties are important to customers of the respective industries and both need to be quantified before they can be monitored or controlled.

The chapter includes an introduction to methods of estimating process variables that cannot be measured directly. This area of work (in which reliable estimates of process variables are obtained from unreliable primary measurements or from measurement of variables on the process periphery) is of ever-increasing importance in large-scale industry.

Two topics important in measurement are treated outside this chapter. They are the effect of dead-time (influenced by siting of sensors) treated in Chapter 6, and the specification of weighing accuracies for a batch process, treated in Chapter 10.

3.1 Process variables — an attempt at classification

Consider a plant that continuously processes a set of raw materials to form a product.

In general, measurements are required of:

(*a*) variables describing the fixed part of the plant;
(*b*) variables describing the outgoing products (allowing feedback control and quality control);
(*c*) variables describing the ingoing raw materials (allowing feedforward control).

By their nature, the variables in class (*a*) are better defined and easier to measure accurately than those connected with the products or raw materials. Raw materials and products both move, often preventing direct sensor installation. Raw materials are frequently inhomogeneous. For outgoing products it is often required to measure ill-defined variables of interest to customers.

Now we can appreciate one of the reasons for the greater precision possible with servomechanisms: they have no product measurement problems to contend with.

Keeping the above discussion in mind we look at the same measurement problem more explicitly. Below we set up three categories of variables according to the ease with which continuous measurement can be made on-site. (The categories necessarily have poorly-defined boundaries.)

(A) *Variables particularly suitable for continuous measurement*
     Examples
     Position, velocity and acceleration (the servomechanism variables)
     Weight and density
     Mechanical stress
     Liquid level
     Gas or liquid pressure or flow rate
     Temperature
     Current and voltage

(B) *Variables for which universally accepted measurement techniques exist but where continuous on-line measurement is more difficult*
     Examples
     Viscosity
     Chemical composition

(C) *Variables that, although economically important, do not have a fully scientific basis for their quantification*
     Examples
     Product colour, taste and crispness (foods)
     Internal cracks, flatness and surface finish (metals)

Although category (A) variables are the easiest to measure, they can still present some formidable problems in practice. In temperature measurement, for instance, a thermocouple measures at only a localised spot, whereas the temperature of interest is usually that in a much larger region. Thermocouples are also often affected by spurious radiation or flame impingement. Radiation measuring thermometers are often affected by scale on the surface of the material being measured or by smoke or steam entering the line of sight.

Moving to category (B), consider chemical composition, which itself contains a very large number of types of application. In general, methods that worked by taking a sample to a laboratory, followed by chemical analysis, could not be extended to continuous on-line use. Instead, a number of physically-based devices, the mass spectrometer being typical, have been developed specifically to satisfy industrial requirements for continuous measurement.

The variables in category (C) are very much the concern of the control engineer since it will often be he who takes a lead in their definition. General guidelines for the definition and subsequent measurement of these variables might be put forward — cautiously, since each problem is unique — as:

(a) Find out what is regarded as important and give it a name if none exists already (for example, 'rollability' of a hot steel ingot).
(b) Find out what can be measured.
(c) See whether (a) can be related to the variables in (b) by a set of consistent curves or, equivalently, by a set of consistent look-up tables.
(d) Convert the curves of (c) into an on-line algorithm remembering that the 'calibration' of (c) was empirical and that it may become invalid if even a non-fundamental change occurs in the process, its operation or its raw material.

The quantification of flatness of steel strip is discussed in Chapter 10.

3.2 Applying sensors to process plant — some practical observations

(a) The scale of operation of most processes brings engineering difficulties that will usually dominate the measurement problem — methods suitable for pilot plants simply cannot withstand the environment of a full-scale production plant.
(b) Even apparently routine measurements may be difficult to achieve reliably. (*Example* — temperature, as discussed in Section 3.1.)
(c) Many processes have unique or near-unique measurement problems. The extent to which these can be overcome will usually determine the ultimate performance that can be achieved by the application of control. (*Example* — flatness of steel strip, as discussed in Chapter 10.)
(d) New methods for the quantification of previously unquantified variables may be required before instrument development can begin.
(e) Errors due to non-homogeneity in the process will usually swamp those due to inherent instrument inaccuracy. (For instance, gas analysis may be in error because of stratification in a large-diameter gas stream.)
(f) Time delays may be influenced by the siting of sensors. A fast-acting sensor that needs to be mounted a long way downstream of a process may be nearly useless for control purposes because of the transport delay introduced.
(g) Often there will be a trade-off between accuracy and speed of response. Very

roughly each measurement can be considered to have error e and time delay τ. For application in a control system, the method chosen should minimise the sum $\alpha|e| + \beta\tau$: α and β chosen to match performance requirements for the controlled systems.

(*h*) Variables that are difficult to measure directly can often be inferred (treated in Section 3.4).

(*i*) Many processes whose fundamental variables cannot be measured directly are being controlled satisfactorily using artificial variables that are combinations of those variables that can be measured relatively easily. (*Example* – control of a reactor on heat-balance calculations, see Section 3.3.)

(*j*) The application of control to a process that was previously uncontrolled often highlights inadequacies in the measurement methods. Because of this a new measuring device cannot be evaluated fully at an uncontrolled plant. Since it is unacceptable to control a plant using an untested sensor, validation at the plant must proceed in two stages:

(i) monitoring without control;

(ii) in a control mode on a trial basis.

(*k*) Experience shows that some measurements that originally were considered indispensable for control can be avoided completely by enforcing repeatability in the process operation and ensuring full information flow on any departure from repeatability.

(*l*) Many variables that are difficult or very costly to measure continuously can be measured at isolated points in time relatively easily (spot measurements), and a satisfactory standard of control still be achieved.

(*m*) Automated inspection to ensure a high standard of quality control is assuming increased importance in nearly every industry. A chronological progression occurs: purely manual visual inspection is first augmented by special instruments whose output is interpreted visually. Computer interpretation of the information is the next logical step. Often such interpretation is a combination of pattern recognition, statistical technique and specialist product knowledge.

(*n*) For those processes where the spatial distribution of a variable must be determined, three approaches are possible:

(i) to use a scanning sensor (discussed below);

(ii) to use a number of fixed sensors and estimate the distribution by curve fitting;

(iii) to use a process model to estimate the distribution.

3.2.1 *Measurement of a distributed variable by a scanning sensor*

Let $x(l, w)$ be a function of the two spatial variables l, representing length and w, representing width, in a product moving in the l direction.

A scanning sensor will follow the dotted line shown in Fig. 3.1. The question naturally arises, how fast must scanning take place in order that successive traverses of the scanner shall yield information useful to the control system? In practice, this means that scanning must be much faster than the rate of change of x in the l direction, i.e. we must have

$$\left|\frac{dw}{dt}\right| \gg \left|\frac{dx}{dl}\right|_{max}$$

Now a situation will often satisfy this requirement initially and a scanning sensor may be fitted and will operate successfully. The next stage of development may be to control the distribution of x in a feedback system driven by the sensor.

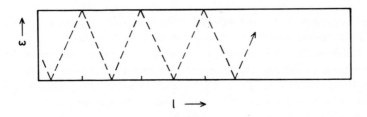

Fig. 3.1 *The track of a scanning sensor on a lengthwise moving product*

A fast-acting system will increase $|dx/dl|_{max}$ and will invalidate the inequality. Hence a scanning sensor must be specified with a sufficient scanning speed to match any control system that is likely to be used. If this is neglected, the situation may arise where the sensor performs perfectly well until closed-loop control is implemented, when the sensor proves inadequate on grounds of scanning speed.

3.3 Control by empirical relations

This type of approach is particularly appropriate where a continuous (as opposed to batch) process is to be maintained in a constant state despite the presence of disturbances. The various measurements that are available are weighted and combined together in a semi-scientific way to produce one or more variables that are then used to drive the process control system. While no process model is needed (and that is one of the advantages of the approach) the semi-scientific combination often makes some basic assumptions connected with mass, energy or chemical balance in the process. For instance, in a continuous chemical process where the internal composition is impossible to measure, the approach might be to produce a single variable from suitable weighted gas flow, gas temperature and gas analysis information. All that is required is to correlate this single artificial variable with the variables of importance. The advantage of this rather vague-sounding method is that it can achieve good results in return for only moderate expenditure of effort. Of course, if the continuous process is allowed to depart far from its desired state, then the method advocated here cannot be expected to achieve very good correction of such large deviations.

3.4 Inferred measurements

Frequently the most important variables in a process cannot be measured directly, at least not on a routine continuous basis. A number of different approaches are then possible. While the ones using on-line models are most attractive when reading a book, they do incur very large expenditures of effort and this is why other, more empirical, methods have also been described.

3.4.1 Direct calculation of the required process variables using simple process relationships, operating on measurements that are available

The principle of the method is illustrated by the following example. In batch process steel-making, the important variable that cannot be measured continuously is the percentage of carbon in the molten steel. However, given a knowledge of the initial percentage at the start of the batch and a measurement of analysis, temperature and flow rate of the gases leaving the process, it is possible to calculate the carbon content $c(t)$ at any time during the batch by use of the relation

$$c(t) = c(0) - \int_0^t \frac{dc}{d\tau} d\tau$$

where $dc/d\tau$ is the rate of removal of carbon with time, calculated from the gas analysis data.

3.4.2 Estimation of process variables from measurements

It is natural to enquire whether computational techniques can be applied to increase the reliability and accuracy of measurement. Some of the problems that might be defined are as follows:

(*a*) Given noisy measurements of the state of a process, produce a best estimate of the true state.

(*b*) Given an exact model of the process and measurements of the output, derive estimates of the internal process variables that cannot be measured directly.

(*c*) Given an approximate model of the process and noisy measurement, derive estimates of the internal process variables.

It can be seen that all three problems could be found in applications with problem (*c*) being quite realistic.

Problem (*a*) is a straightforward filtering problem. It will be discussed only briefly. Problem (*b*) is the so-called *observer problem*. The solution involves using the process model to produce a reconstructed state vector. Problem (*c*) can be solved completely under very restrictive conditions (linearity, independence, Gaussian statistics) using the Kalman filter. However, as we shall indicate, with perseverance and engineering compromise, the Kalman filter can be made to work very satisfactorily in industrial situations.

The Kalman filter is an algorithm that allows optimal estimation of the internal variables of a process from noisy measurement. It has to contain a dynamic model of the process and needs to be given a quantitative description of the corrupting noises. The derivation of the filter can be found in the original references (Kalman, 1960, Kalman and Bucy, 1961). A heuristic explanation of the principle is given here in Section 3.7. Applications of the filter are described in Davies (1978), Litchfield *et al.* (1979), Page (1979) and Wick (1978).

3.5 The Kalman filter − principle of operation

From our point of view, the Kalman filter is a realistic tool that allows reliable estimates of unmeasurable process variables to be made on-line. Such estimates can then be used for closed-loop control or for operator guidance.

The filter also allows inaccurate measurements to be processed to give a more accurate data set on which to base control actions.

Finally, the filter can estimate unknown process parameters.

When it is seen that the filter can achieve all three objects simultaneously, its usefulness is obvious.

It has to be said, however, that the filter algorithm is complex, there is a need for a good process model and each application needs a considerable input of development effort before it performs as required. Many problems have been solved, however, and a methodology for application is beginning to emerge.

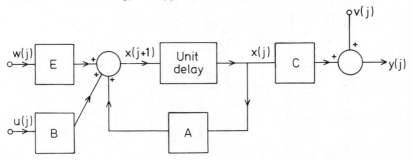

Fig. 3.2   *The linear discrete time process model with noise inputs* w, v *required as a starting point in the application of the Kalman filter.* x *is the system state: it is to be estimated from measurement of the noise corrupted output* y

The problem is defined in terms of Fig. 3.2. The internal state x of a process is to be estimated from available measurements y. The filter is connected to the process as shown in Fig. 3.3. The user is required to input two matrices Q, R that are statistical measures of the errors in the model and in the measurement respectively. The filter then iterates in real time in a feedback-loop mode to modify its estimate $\hat{x}$ of the process state until the estimation error $x - \hat{x}$ is minimised. The gains in this feedback loop are given by a matrix K, the so-called Kalman gain matrix. K is a set of time-varying gains and is calculated and reset at each time step in the algorithm.

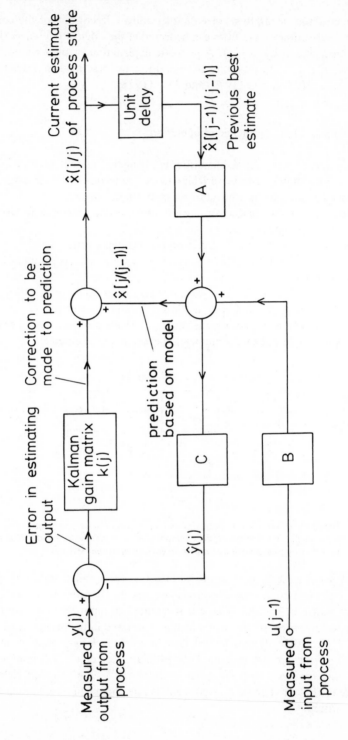

Fig. 3.3   *The Kalman filter. It receives process inputs and outputs, contains an internal process model and outputs real-time estimates of the process states*

A heuristic derivation of the Kalman filter algorithm is given in Section 3.7. The principal problems to be overcome before it can be applied on-line are:

(*a*) The matrices Q and R need to be estimated, either from available data or by other techniques.

(*b*) A process model of sufficient accuracy and speed needs to be available. Often only a subset of the process needs to be modelled in detail.

(*c*) The extent to which assumptions of process linearity and noise whiteness are satisfied needs to be investigated and appropriate extensions to the algorithm incorporated, should they be needed.

3.6 The Kalman filter – illustrative examples

An example from the growing number of successful applications reported is that of estimating the core temperature of steel ingots, based on the measurements that are available processed by a Kalman filter. The example (Wick, 1978) was chosen because it is operating on a production plant.

Figures 3.4 and 3.5 illustrate the main points of the application. A steel ingot is placed in the furnace at time zero. As a nominal initial condition, it is assumed that the core temperature is $150°C$ higher than the surface temperature (this arises because the ingot has recently been cast from liquid). The filter then operates in real-time on the available furnace measurements to yield a continuous estimate of core temperature. This key parameter is displayed to the furnace operator who uses it as a criterion of uniformity of heating.

3.6.1 Discussion of a second application example

In steel-making by the Basic Oxygen Process, the temperature and composition of the steel, although of vital importance, cannot be measured directly and have to be inferred from variables that can be measured. Attempts to use the Kalman filter operating on these available measurements, which consisted largely of measurements of variables in the waste gas from the process, proved to be of only limited success. However, success was achieved once a continuous measurement of the total process weight was added to the measurement vector y. Although the process weight measurement had not been considered of much significance until that time, it was a very accurate noise-free measurement. With the incorporation of process weight into the algorithm, the estimation accuracy for all variables in the state vector was improved quite markedly, to the point where the algorithm could be considered to be 'commercially' valuable.

3.7 Heuristic derivation of the Kalman filter for a linear discrete-time process

$$x(j+1) = Ax(j) + Bu(j) + Ew(j),$$
$$y(j+1) = Cx(j+1) + v(j+1). \qquad (3.1)$$

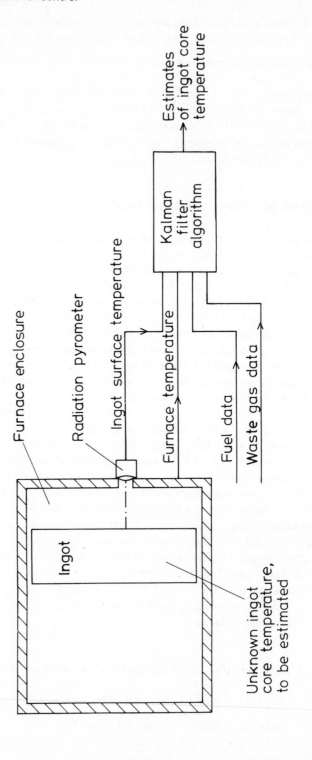

Fig. 3.4 *Application of the Kalman filter to the estimation of ingot core temperature*

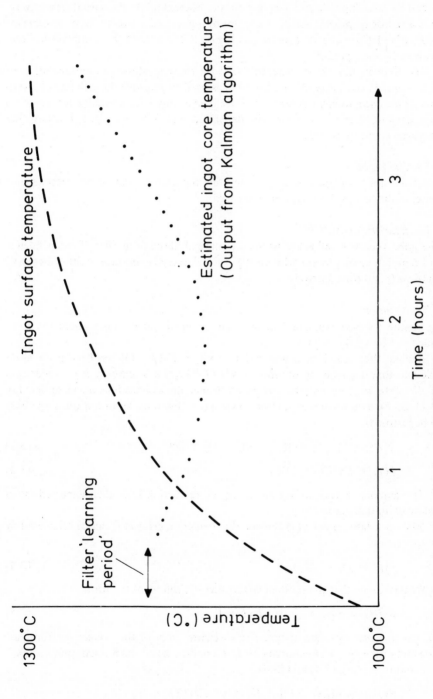

Fig. 3.5   *Sketch of actual results obtained by Wick on a production ingot heating furnace*

Equation (3.1) represents a linear process with time-invariant parameters disturbed by noise inputs v and w acting on the process and on the output respectively. (x is a vector of process states, u a vector of inputs and y a vector of measurement. A, B, C and E are constant matrices.) Figure 3.2 illustrates the configuration represented by the equation.

The Kalman filter is, in general, a tool for the smoothing and extrapolation of time series. In our applications, the filter is used to improve the estimate of a variable whose measurement is noisy or to yield an estimate for a variable for which no measurement of any sort exists. The derivation that follows is slanted towards the envisaged applications area.

3.7.1 The problem

Given the noise-corrupted data, y, determine the best estimate $\hat{x}$, in a least-squares sense, of the process internal state vector x.

3.7.2 Heuristic derivation

The noise vectors v and w are assumed to have idealised properties of independence and complete randomness ('whiteness'). Their covariance matrices are the (diagonal) matrices Q and R respectively.

3.7.3 Notation

We denote the best estimate $\hat{x}$ of the state at time j, based on measurements up to time $j - 1$ by $\hat{x}[j/(j-1)]$.

Assume that somehow a best estimate $\hat{x}[(j-1)/(j-1)]$ has been made available and that it is required to estimate $\hat{x}[j/(j-1)]$, i.e. a one-step ahead prediction of the state and the output is required. Under the assumptions made earlier, Eq. (3.1) can be used for the prediction to yield (in a rigorous definition this step needs to be justified):

$$\hat{x}[j/(j-1)] \; = \; A\hat{x}[(j-1)/(j-1)] + Bu(j), \tag{3.2}$$

$$\hat{y}(j) = C\hat{x}[j/(j-1)]. \tag{3.3}$$

The random vectors do not appear in the prediction since their expected values are by definition zero.

When a measurement $y(j)$ is made the prediction error $\tilde{y}(j)$ can be calculated by an expression

$$\tilde{y}(j) = y(j) - \hat{y}(j). \tag{3.4}$$

We now argue that the optimal estimate $\hat{x}(j/j)$ will be of the form

$$\hat{x}(j/j) = \hat{x}[j/(j-1)] + Ky(j), \tag{3.5}$$

i.e. that the best estimate at time j is a weighted sum of the prediction from past data and of the current measurement. K is a matrix called the Kalman gain matrix.

From Eqs. (3.3), (3.4) and (3.5):

$$\hat{x}(j/j) = \hat{x}[j/(j-1)] + K\{y(j) - C\hat{x}[j/(j-1)]\}$$

and substituting from Eq. (3.2):

$$\hat{x}(j/j) = A\hat{x}[(j-1)/(j-1)] + Bu(j) + K\{y(j) - C\hat{x}[j/(j-1)]\}. \quad (3.6)$$

Equation (3.6) is the state estimation algorithm. The Kalman gain matrix needs to be calculated before it can be used. (In the noise-free situation. Eq. (3.6) represents the Luenberger observer; it can be shown to converge to the correct value for all finite values of the gain matrix K.)

The *state estimation error* is defined

$$\tilde{x}(j) = x(j) - \hat{x}(j/j). \quad (3.7)$$

From Eqs. (3.1), (3.6) and (3.7)

$$\tilde{x}(j) = (I - KC)[A\tilde{x}(j-1) + Ew(j-1)] - Kv(j), \quad (3.*)$$

where I is the identity matrix.

A covariance matrix $P(j)$ is defined as

$$P(j) = \mathscr{E}\{\tilde{x}(j)\tilde{x}(j)^T\}, \quad (3.8)$$

where $\mathscr{E}$ indicates expected value. P is a matrix that indicates the 'goodness' of the state estimation.

Equation (3.7) is substituted into Eq. (3.9) and it yields a quadratic in the matrix K which can be solved after some manipulation to yield, together with earlier equations, the final algorithm:

$$P(j) = [I - K(j)C]P^*(j)$$

where

$$P^*(j) = AP(j-1)A^T + EQE^T,$$

$$K(j) = P^*(j)C^T[CP^*(j)C^T + R]^{-1},$$

$$\hat{x}(j/j) = [I - K(j)C]\{A\hat{x}[(j-1)/(j-1)] + Bu(j-1)\} + K(j)y(j).$$

The equation for $\hat{x}$ is the on-line estimator. It makes use of the three equations above but they require no process data input and so can be pre-computed if necessary.

For the example treated here (linear, time-invariant process, noise assumptions satisfied) the matrices P and K tend to a steady state as j increases.

Figure 3.3 shows a block diagram of the filter algorithm.

When, as is often the case, the noise characteristics fail to satisfy the whiteness assumption, the situation may still be handled by postulating a fictitious white noise source feeding into a noise model. This noise model is then considered to be part of the process and the enhancement allows the original assumptions to be satisfied.

Non-linear processes can be treated by using the so-called extended Kalman filter. This involves linearisation at every time step but otherwise the technique is the same as before. Note, however, that the amount of computation is very considerably increased.

The extended Kalman filter can be used to estimate both the state and the parameters of a process. The procedure is to designate the parameters as dummy states (the resulting product terms introduce non-linearities, hence the need for the extended version of the Kalman filter).

Chapter 4

Simple controllers and methods for setting their parameters

4.0 Introduction

This chapter is concerned with the application of traditional 'off-the-shelf' controllers. Even complex industrial processes often can be controlled adequately by such controllers whose attraction lies in their ready availability, ease of installation and simple maintenance. Controller commissioning consists in allocating numerical values to a few parameters. A number of different methods, starting with the purely empirical, are put forward for selecting the parameters.

4.1 Types of controller

4.1.1 On–off controllers
A simple on–off controller can be realised by mounting a flag on the pointer of a moving-coil indicator. The flag interrupts a beam of light and so operates a photo-relay. An on–off controller will often be the correct choice for application to a small thermal process such as a laboratory oven.

4.1.2 Three-position controllers
These controllers switch the output between three conditions: high, low, and off. Their simplicity implies simplicity in the actuators that they drive – an important economic factor. Thus, they find wide application in practice.

A typical application is to the control of a large coal-fired furnace. Coal is fed to the furnace along a worm that can rotate at two different speeds, selectable by a remotely-controlled gearbox. When the furnace temperature is below the desired value, the worm drives at its higher rate, reducing to its lower rate once the desired value is exceeded. Should the furnace temperature exceed a user-fixed upper limit the worm is stopped. Figure 4.1 illustrates the operation. Notice in particular the simple actuator arrangement. A two-speed gearbox is all that is required. A controller with a continuous output would have required an infinitely variable-speed drive for the worm, with consequent increase in cost.

4.1.3 One-, two- and three-term controllers

From a theoretical standpoint, the simplest controller of all is one that compares a measured value with the desired value and, multiplying by a gain k, closes the control loop. Such a controller is called a *one-term controller* or a *P controller* (P for proportional).

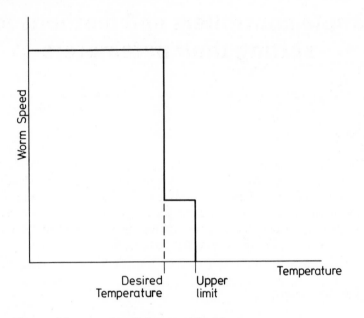

Fig. 4.1   *Three-position control applied to a coal-fired furnace*

Any discussion of steady-state accuracy introduces the desirability of having integration present in the control loop. A controller with a transfer function $[k + (1/T_I s)]$ is called a *two-term controller* or a *PI controller*.

Finally, an ability to control the rate and extent of oscillations of a controlled system requires a derivative term in the controller. So we arrive at the *three-term controller* or *PID controller* with transfer function $[k + (1/T_I s) + T_D s]$ (see Fig. 4.2).

Controllers in these categories have been around for many years — considerably pre-dating the electronic era — even so they should not be despised. Although no longer mentioned in the respectable literature, they are even now controlling and safe-guarding thousands of processes.

4.2 Approaches to the control of a single-input/single-output process

(*a*) (Completely empirical.) Given a suitable measuring sensor and final control element, close the loop through an appropriate commercial controller and adjust the controller gain and other controller strategy by rules of thumb to obtain an acceptable performance.

If accompanied by sound process knowledge and engineering sense, such an approach will be perfectly acceptable for many straightforward applications.

Suitable tuning methods for PID controllers are suggested in Sections 4.3—4.5.

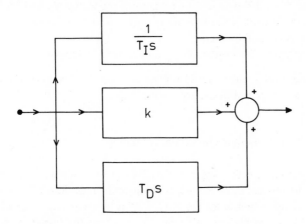

Fig. 4.2 *A three-term controller*

(*b*) For many processes, it will be impossible because of the importance and nature of the process to indulge in major on-process tuning. However, it will often be possible to obtain an open-loop step response and to fit a simple model by graphical or numerical techniques to these data.

Methods exist for rapidly determining controller settings, given an approximate process model obtained in this way (see Section 4.4).

(*c*) A full understanding of the dynamic characteristics of the process requires a detailed knowledge of the physical structure of the process and an accurate mathematical model of each sub-system that exists within that structure. Such a knowledge will not be obtained without expenditure of very considerable effort. However, given such quantitative understanding, the full weight of control techniques, in particular the root-locus technique, can be brought into play. The effects of disturbances, both periodic and stochastic, of non-linearities and parameter drift can be investigated. The process, complete with its proposed control system, can be simulated and changes to the process itself or to its mode of operation will be suggested to management at this stage, with great benefits, in the light of the concentrated knowledge then existing. Once the control system has been installed, there will be a store of knowledge available for its evolution and further development.

4.3 On-plant tuning of three-term controllers

The simplest method of tuning three-term controllers directly on plant is as follows:

(*a*) Close the control loop through the controller and with derivative and integral actions out of operation increase the controller gain until continuous oscillation results. Note the period of this oscillation, denoted by (say) T^*.

(*b*) Set the derivative time T_D and the integral time T_I such that $T_D = T_I = T^*/2$.

(*c*) Observe the period of oscillation of the resulting system. The aim is to keep this period at T^* and this would be achieved if procedure (*b*) had been carried out perfectly. If adjustment is needed, the period should be increased by reducing T_D and should be decreased by increasing T_I.

(*d*) Adjust the controller gain until the desired decay rate is obtained.

For a PI controller, the procedure is to set $T_I = T^*/2$, which should result in a 50% increase in the period of the controlled loop.

4.4 Ziegler–Nichols methods for controller tuning

A second method, due to Ziegler and Nichols (1942) requires the open-loop step response of the process to be recorded. The process, regardless of its actual order, is then modelled by a first-order lag $K/(1 + sT_1)$ in series with a dead-time e^{-sT_2}. K, T_1 and T_2 are simply read off from the graph of the process step response as shown in Fig. 4.3(*a*).

Of course, not every process can be modelled by this type of expression:

(*a*) Processes containing integrators would be modelled equivalently by an expression of the form Ke^{-sT}/s [Fig. 4.3(*b*)].

(*b*) Processes with oscillating step response would need at least a second-order model to represent them.

(*c*) Processes with the step response as shown in Fig. 4.3(*c*) need at least a second-order model and a numerator term if they are to be modelled at all accurately.

Once the process model of form $(Ke^{-sT_2})/(1 + sT_1)$ is available, Ziegler and Nichols (1942) suggest the following settings for the three-term controller:

$$\text{Proportional gain only}\quad k = \frac{1}{K}\frac{T_1}{T_2};$$

$$\text{Proportional + integral controller}\quad k = \frac{0{\cdot}9}{K}\frac{T_1}{T_2},\quad T_I = 3{\cdot}3T_2;$$

$$\text{Three-term controller}\quad k = \frac{1{\cdot}2}{K}\frac{T_1}{T_2},\quad T_I = 2T_2,\quad T_D = 0{\cdot}5T_2.$$

Notice that the settings suggested by Ziegler and other authors envisage step changes in desired value. With settings chosen in this way, the response to step changes in load will tend to be too oscillatory. In situations where the main requirement is to compensate for load changes rather than desired value changes, somewhat different settings are required (notice that it is only because of the integral term that this occurs). The tuning of controllers for disturbances in load is considered by Smith (1972, Chapter 6).

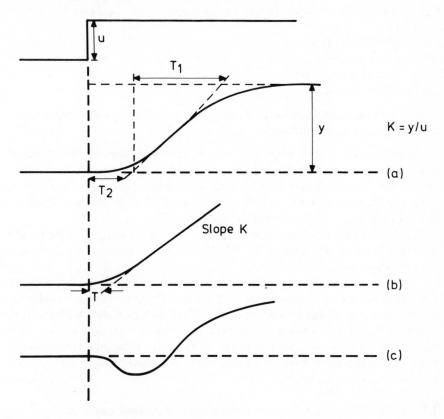

Fig. 4.3   *Open-loop process step responses*

Ziegler and Nichols suggest an alternative method. With the process in closed loop, but with no derivative action, find the gain k^* that results in steady oscillation at some period, say T^*. Recommended controller settings are then:

    For the proportional only controller $k = 0.5k^*$;

    For the PI controller $k = 0.45k^*$, $T_I = 0.83T^*$;

    For the PID controller $k = 0.6k^*$, $T_I = 0.5T^*$, $T_D = 0.125T^*$.

4.5 Tuning methods relying on possession of a process model

4.5.1 Interpreting the results of simple tests to obtain a process model
A general process model can be considered to be made up of:

(a) Dead-time, modelled in the form e^{-sT_d} where T_d is the dead-time.

(b) One or more integrators, of the form $1/s$.

(c) Dynamic elements of the form $1/(1 + sT_1)(1 + sT_2) \ldots (1 + sT_n)$. (Repeated factors, irreducible quadratic factors and numerator terms might also be encountered.)

(d) Time-invariant gain K.

(e) Nonlinear elements.

(f) Time-varying elements.

(g) Stochastic features.

It is important to understand the simpler aspects before embarking on time-consuming complex investigations.

(a) Dead-time T_d, at least where it is constant, can be determined from a straight-forward step test conducted on the (open-loop) process. Let a process that consists only of dead-time T_d be put in series with a manipulable gain k in closed-loop. At some gain k^* the system will execute steady sinusoidal oscillations with frequency ω^*. Let the period of these oscillations be denoted T^*.

For continuous oscillation, there must be $-180°$ phase shift across the process, from which it can be seen that $T_d = T^*/2$, i.e. the oscillations have a period of twice the dead-time.

(b) If the process can be represented by a dead-time in series with an integrator then, since the integrator contributes $-90°$ phase shift at every frequency when the process oscillates in closed-loop, the period T^* of such oscillation will satisfy the equation $T_d = T^*/4$. Knowledge of these and similar relations, which can be worked out for each case, allow estimates to be made of the approximate transfer function of a process in a very short time from the results of simple tests that most process managers will tolerate.

4.5.2 Using the process model to determine the best controller settings

Given a process model, obtained as in Section 4.5.1, the most direct route to finding controller coefficients is to use an iterative numerical technique.

First a criterion, J, is defined that is to be minimised by choice of controller co-efficients. Possible choices are:

(a) $J_1 = \int_0^\infty |e(t)| \, dt$;

(b) $J_2 = \int_0^\infty e(t)^2 \, dt$;

(c) $J_3 = \int_0^\infty t|e(t)| \, dt$;

where it is understood that at time 0 a step is applied to the system, either to the input or as a load disturbance. $e(t)$ is the error between actual value and desired value.

A computer program is written in which the process model is simulated in series with a model of the controller in closed loop. Arbitrary parameters are first allocated to the controller. A simulated step is applied and J is calculated. Now using a hill-climbing algorithm, the controller coefficients are manipulated iteratively. At each iteration, J is reduced and when it cannot be reduced further, the controller parameters in the model are taken as the best possible. Ready made programs exist for hill-climbing. See Leigh (1980) for a summary of suitable methods.

4.6 Comparison of tuning methods

Lopez *et al.* (1969) and Rovira *et al.* (1969) use hill-climbing techniques to minimise error criteria. Lopez assumes step changes in desired value whereas Rovira assumes step changes in load. A method due to Martin *et al.* (1977) operates conceptually in the complex plane. The integral term in the controller is chosen to compensate for the dominant pole in the process transfer function and the derivative term is chosen to compensate for the next most significant pole. The controller gain is then adjusted to give the desired step response characteristic. Of the three methods, this offers the advantage that it could easily be incorporated into a simple adaptive system, since the controller gain is independent of the other parameters and can be adjusted easily to take account of changes in the process.

Martin *et al.* (1977) compared four methods (Ziegler—Nichols, Lopez, Rovira and Martin) using results from a simulated chemical reactor. His results show rather oscillatory responses for the Ziegler—Nichols settings. He concluded that the Rovira and the Martin method performed best in the tests he applied.

4.7 Commercially available three-term controllers

Looking rather conventional and without necessarily being microprocessor-based, currently available three-term controllers are, in their more sophisticated versions, extremely versatile. As an example, we summarise below the specification of the Hartmann and Braun controller type *SK*. (For complete particulars see Hartmann and Braun data sheet 62—4.10 en.)

SK controller brief specification: One-, two- or three-term control using plug-in modules.

Inputs: Voltages, currents or temperature signals from thermocouples or resistance thermometers. Thermocouple break protection, cold-junction compensation, linearisation.

Controlled variable: Displayed by a stepping-motor-driven contactless (capacitive) servo.

Desired value: Set locally or remotely with bumpless transfer between. Desired value is stored in a 10-bit register.

Controller: Adjustable output limits built in. Integration has anti-wind-up facility. The differentiator operates on the error, on the output or on an auxiliary variable. Asymmetrical derivative action (acting on only positive or negative going signals) available.

Output: Digital output memory allows bumpless transfer between modes.

Alarms: Non-contact pick-ups visible on the indicator scale for controlled variable alarms.
Internal deviation alarms.

Computer connection: Direct digital control or supervisory control. Communication is through pulse amplitude modulated, pulse duration modulated or pulse code modulated computer signals.
Transfer logic circuits are used for automatic transfer of the operating mode.

DDC back-up: Bumpless transfer into a chosen safety mode is achieved in the event of computer failure. (Similar safeguards are available against computer failure while in the supervisory control mode.)

Multiplexing: Available to reduce the wiring to the computer in large systems.

Chapter 5

Direct digital control (DDC) algorithms for single-input / single-output processes

5.0 Introduction

When control is to be implemented in a digital computer, one way forward is to use a sampled-data (discrete-time) version of a well-established analogue controller, such as a three-term controller. Alternatively, a specifically discrete-time controller can be synthesised that (like the dead-beat controller) has no analogue counterpart.

Both approaches are considered in this chapter, as are questions regarding choice of sampling interval, computational time and memory requirements of the algorithms.

5.1 A selection of DDC algorithms

5.1.1 PID algorithm
A simple DDC algorithm for a single-input/single-output system is a digital realisation of the three-term analogue controller of transfer function

$$G(s) = k\left(1 + \frac{1}{T_1 s} + T_2 s\right) \tag{5.1}$$

in which the three coefficients can be determined by a number of methods from classical control theory or by the Ziegler–Nichols semi-empirical method (Ziegler and Nichols, 1942).

Using simple approximations to differentiation and integration the discrete-time form of the three-term controller is

$$u_n = k\left(e_n + \frac{T}{T_D}\sum_{i=0}^{n} e_i + \frac{T_I}{T}(e_n - e_{n-1})\right) + u_0. \tag{5.2}$$

u_n is supposed to be the position of a process actuator. Alternatively, the algorithm can be expressed

$$u_n = k\left((e_n - e_{n-1}) + \frac{T}{T_D}e_n + \frac{T_I}{T}(e_n - 2e_{n-1} + e_{n-2})\right), \tag{5.3}$$

which is an incremental form.

Equation (5.2) can cause undesirable disturbances when it is first switched on since the actuator will be moved instantaneously to the calculated value from wherever it happens to be. Some provision needs to be built in to avoid difficulties from this source.

The algorithms are often modified to prevent step changes in desired value from being differentiated.

Algorithm (5.2) becomes

$$u_n = k\left(e_n + \frac{T}{T_D}\sum_{i=0}^{n}e_i - \frac{T_I}{T}(y_n - y_{n-1})\right) + u_0 \tag{5.4}$$

y_n being the process output.

A further modification is to incorporate a facility to prevent 'wind-up' of the integral term. The simplest method is to disable the integration when the actuator becomes saturated. The second version of the algorithm (5.3) implicitly contains this facility. Similarly some mechanism is needed to prevent the integral term building up to a large spurious value after a set-point change. The simplest strategy is to immobilise the integral term completely whenever the input to the controller exceeds some pre-set value. The controller output must not exceed the range that can be accepted by the actuators on the plant. This requirement is easily built into a program.

The three-term controller algorithm can be represented in the Z domain by the discrete-time transfer function

$$G(z) = \frac{a_0 + a_1 z^{-1} + a_2 z^{-2}}{1 + b_1 z^{-1} + b_2 z^{-2}}, \tag{5.5}$$

where the coefficients a_i, b_i depend on the coefficients in Eq. (5.2) and on the sampling interval. (The relations between the coefficients of Eqs. (5.1) and (5.5) are obtained easily by using difference relations to approximate differentiation and summation relations to approximate integration.)

Optimum settings for the coefficients of discrete-time three-term controllers have been calculated numerically by Lopez *et al.* (1969) and the results expressed as curves.

The curves are particularly useful in allowing a quantitative appreciation of the interrelation between process parameters, sampling rate, controller settings and the type of response resulting.

5.1.2 *Cascade control algorithm*

This consists typically of a PID control loop with an inner PI loop. It can be realised by the equations:

$$w_1(k) = w_1(k-1) + [r(k) - y_1(k)],$$

$$u_1(k) = k_{p_1}(r(k) - y_1(k)) + k_{i_1}w_1(k) - k_{d_1}(y_1(k) - y_1(k-1)),$$

$$e_2(k) = u_1(k) - y_2(k),$$

$$w_2(k) = w_2(k-1) + e_2(k),$$

$$u_2(k) = k_{p_2} e_2(k) + k_{i_2} w_2(k).$$

The suffixes 1 and 2 refer to the outer and inner loops respectively. $r(k)$ is the desired value at time step k. The best values for the coefficients are preferably found by a simulation exercise, since the number of parameters makes on-line tuning difficult.

5.1.3 Finite-settling time control

A sampled data system is said to be finite-time settling if its response can be expressed as a finite polynomial in the operator z. This can be achieved provided that the poles of the system are at $z = 0$.

A control algorithm, based on this approach, is described below. The aim of the algorithm is to obtain a step response similar to that of Fig. 5.1. This is often called *dead-beat* control.

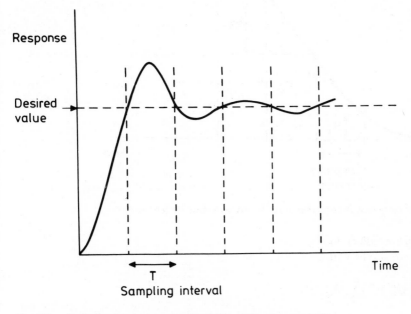

Fig. 5.1 *Desired process response with finite settling-time controller*

The method (Auslander *et al.*, 1978; Takahashi *et al.*, 1975) aims to achieve a high level of control while at the same time making little demand for process knowledge or complexity of on-line computation.

The starting point is the open-loop step response curve of the process which is assumed to be of the form shown in Fig. 5.2.

The discrete-time transfer function of the process can be written in the form

$$\frac{Y(z)}{U(z)} = G(z) = \sum_{i=1}^{\infty} g_i z^{-i}.$$

Let the sequence $\{g_i\}$, $i = n, \ldots$ be approximated by an exponentially decaying sequence. Then, approximately,

$$G(z) = \sum_{i=1}^{n-1} g_i z^{-i} + \frac{g_n z^{-n}}{1 - pz^{-1}},$$

in which the parameter p governs the rate of decay of the sequence $\{g_i\}$, $i = n, \ldots$.

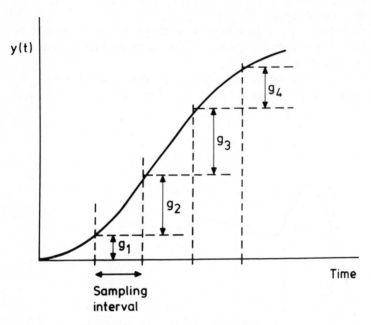

Fig. 5.2   *Open-loop process step response used as a basis for the model*

The g_i are defined in Fig. 5.2.
$G(z)$ can also be written

$$G(z) = \sum_{i=1}^{n} b_i z^{-i} / (1 - pz^{-1})$$

where

$$b_1 = g_1$$

$$b_i = g_i - pg_{i-1}, \quad i = 2, \ldots, n.$$

p, a factor between 0 and 1, will usually be chosen so that the steady-state value of $y(t)$ given by the model agrees with the measured value. This yields a formula for p,

$$p = 1 - \left[g_n \bigg/ \left(k_p - \sum_{i=1}^{n-1} g_i \right) \right],$$

in which k_p is the steady-state gain of the process.

Equivalent state-space model – which will be found useful later,

$$x(k+1) = Px(k) + Qu(k),$$

$$y(k) = Cx(k),$$

in which

$$P = \begin{bmatrix} 0 & 1 & 0 & \ldots & 0 \\ 0 & 0 & 1 & \ldots & 0 \\ \cdot & \cdot & \cdot & \ldots & \cdot \\ \cdot & \cdot & \cdot & \ldots & \cdot \\ \cdot & \cdot & \cdot & \ldots & \cdot \\ 0 & 0 & 0 & \ldots & p \end{bmatrix},$$

$$Q = \begin{bmatrix} g_1 \\ g_2 \\ \cdot \\ \cdot \\ \cdot \\ g_n \end{bmatrix}, \quad C = (1, 0, \ldots, 0).$$

The finite settling-time controller with input consisting of the process output $y(k)$ and the process state vector $x(k)$ then can be shown to take the form:

$$u(k) = k_I \sum_{i=0}^{k} [r(i) - y(i)] - \sum_{j=1}^{n} k_j x_j(k),$$

in which

$$k_I = 1 \bigg/ \left(\sum_{i=1}^{n} b_i \right),$$

$$k_1 = 0, \quad k_2 = k_3 = \ldots = k_{n-1} = k_I,$$

$$k_n = \left[1 + p - \left(\sum_{i=1}^{n-1} g_i / k_I \right) \right] \bigg/ g_n.$$

See Takahashi *et al.* (1978) for the details of the derivation.

The implementation of the algorithm is shown in Fig. 5.3.

The chief practical disadvantage of the method relates to the availability of the process state vector x. Usually, this will not be available and a practical method must be found to allow the vector $x(k)$ to be approximated.

For the estimation of $x(k)$, an appropriately simple extension to the algorithm has been suggested.

$$\hat{x}^0(k+1) = P\hat{x}(k) + Qu(k), \quad \hat{x}^0(0) = 0,$$

$$\hat{x}(k+1) = \hat{x}^0(k+1) + f[y(k+1)] - C\hat{x}^0(k+1),$$

in which

$\hat{x}^0(k)$ is an *a priori* estimate of $x(k)$,

$\hat{x}(k)$ is the final estimate of $x(k)$,

f, in the absence of unmeasurable disturbances, can be obtained from $f^T = (1, p, p^2 \ldots p^{n-1})$.

Q and C are obtained from the state-space model.

Of course, if noise problems are serious, the filter gain f can be chosen using Kalman filter techniques (Chapter 3) but then the simplicity of the whole operation involved in applying the algorithm is lost.

Note that if the open-loop response of the process is oscillatory the approach, somewhat modified, can still be applied (see Tomizuka *et al.*, 1977).

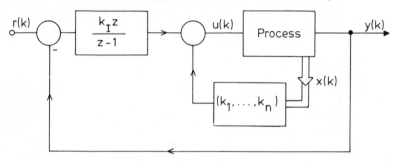

Fig. 5.3 *The finite settling-time controller implemented*

The basic algorithm has been modified by Dahlin (1968) and by Kalman (Chiu *et al.*, 1973).

Dahlin's aim was to achieve a smoother step response, but his algorithm has been found (Uronen and Yliniemi, 1977) to be very sensitive to process parameter variations. Kalman's modification aims to achieve minimum settling time.

5.1.4 Compensator algorithm
This algorithm consists of a digital realisation of a compensating transfer function such that the poles of the closed-loop system are positioned in specified locations. Such an approach pre-supposes that a model of the process is available, when the coefficients of the controller are found by solution of simultaneous equations (see also Section 5.2).

5.1.5 Model following adaptive control
This type of algorithm is particularly suited to the control of time-varying processes. It consists of two parts; identification and automatic controller parameter selection.

5.1.6 Half proportional control

This nonlinear controller is described by the relations:

$$u(k) \; = \; u(k-1) + \Delta u(k),$$

$$\Delta e(k) \; = \; e(k) - e(k-1),$$

$$\Delta u(k) \; = \; c_1 \Delta e(k) \quad \text{if } |e(k)| - |e(k-1)| > 0,$$

$$= \; 0 \qquad\qquad \text{if } |e(k)| - |e(k-1)| < 0,$$

$$= \; c_2 \{e(k) - P \operatorname{sign} [e(k)]\}$$

$$\text{if } e(k) = e(k-1) = \ldots = e(k-N) \text{ and } |e(k)| > P.$$

c_1 is the main gain of the controller. c_2, P and N play a subsidiary role.

5.1.7 Optimal state feedback using an observer

The states are estimated by Luenberger observer and optimal state feedback is calculated by solution of the matrix Riccati equation.

5.1.8 Self-tuning control

Another form of adaptive control which is described in Section 7.1.

5.1.9 Combined feedforward–feedback controller

Let V be a measurable process disturbance. The algorithm is

$$u(k) \; = \; -\frac{P_1(z^{-1})}{P_2(z^{-1})}e(k) + \frac{P_3(z^{-1})}{P_4(z^{-1})} V(k),$$

where the P_i are polynomials in z^{-1}. Often a simple lead-lag filter is used for feedforward. This can be realised through the arrangement:

$$\frac{P_3(z^{-1})}{P_4(z^{-1})} \; = \; \frac{a_0 + a_1 z^{-1}}{1 + b_1 z^{-1}},$$

5.2 Controller design in the z-domain

Let the pulse transfer function of the process, together with the sample and hold device, be $G'(z)$ (see Fig. 5.4). Let $H(z)$ be the desired pulse transfer function of the closed-loop system; then since

$$y(z) \; = \; G'(z)C(z)[v(z) - y(z)].$$

Then

$$C(z) \; = \; \frac{1}{G'(z)} \frac{y(z)}{v(z) - y(z)}$$

and therefore if we choose $C(z)$ to satisfy

$$C(z) = \frac{1}{G'(z)} \frac{H(z)}{1 - H(z)}$$

the desired closed-loop performance will result. (Note that $H(z)$ must contain delays equivalent to those in $G'(z)$, otherwise $C(z)$ will not be physically realisable.) $G'(z)$ can be derived from the simple models of the previous section by making use of the relation:

$$Z\left\{\frac{1 - e^{-sT}}{s} G(s)\right\} = (1 - z^{-1}) Z\left\{\frac{1}{s} G(s)\right\},$$

which is useful since the term $(1 - e^{-sT})/s$ represents the sample and hold device which will be present in practical control loops. If the simple Ziegler model $G(s) = Ke^{-sT_2}/(1 + sT_1)$ (see Section 4.4) is being used to represent the process, then the overall pulse transform for the model including the sample and hold device is given by

$$G'(z) = K \frac{(1 - z^{-1})z^{(1-T_2/T)}(1 - e^{-T/T_1})}{(z - 1)(z - e^{-T/T_1})}$$

And this expression, in conjunction with the equation for C, can be used as the basis for controller design, so that (see Fig. 5.4):

$$C(z) = \frac{H(z)}{1 - H(z)} \frac{(z - 1)(z - e^{-T/T_1})}{K(1 - z^{-1})z^{(1-T_2/T)}(1 - e^{-T/T_1})}$$

where $H(z)$ is the desired pulse transfer function of the closed-loop system and T is the sampling interval.

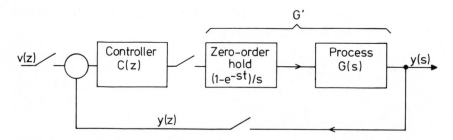

Fig. 5.4   *Synthesis of a controller in the z domain*

In selecting the desired transfer function, H, it must be borne in mind that:

(*a*) If H is set equal to unity, division by zero occurs and the controller does not exist.

(*b*) If the process contains a transport delay, then H must contain a delay at least equal to this. Otherwise the controller will contain positive powers of z and be un-realisable since this implies prediction.

(*c*) H must not be such that a non-minimum phase singularity (a pole or zero in the right half s-plane or outside the unit circle in the z-plane) is cancelled. For instance, if a zero outside the unit circle in the z-plane is to be cancelled, this necessitates a controller with a pole outside the unit circle and leads to instability.

(*d*) Sampled data controllers for processes whose characteristics are not far removed from the non-minimum phase region can experience an undesirable phenomenon that has been called 'ringing', which leads to excessive actuator oscillation. Techniques for the elimination of ringing amount to the removal of critical poles from the controller. See Chiu *et al.* (1973) for details and examples.

Further techniques for tuning discrete controllers are given in Chiu *et al.* (1973) where the algorithms of Dahlin and Kalman are shown to be reducible to PID form. Thus, the Dahlin and Kalman methods lead to techniques for the specification of the PID controller coefficients.

In specifying sampling-time it is important to remember that if a sudden disturbance to a controlled process occurs, no correction can be made until the next scan takes place. For certain processes, this consideration will be critical in deciding the longest sampling interval that can satisfactorily be used.

5.3 Comparison of DDC algorithms

In designing a scheme, a decision has to be made on which algorithms to use. Many papers include simulation studies illustrating the apparent superiority of the more complex algorithms. However, experience with real implementations on plant does not give quite such an optimistic picture. Unbehauen *et al.* (1976) compared seven DDC algorithms by implementing them to control a pilot scale heat exchanger. The seven algorithms were those detailed in Sections 5.1.1–5.1.7. Unbehauen's main conclusions confirm that the PID and cascade controllers give good results in situations where step disturbances are encountered and that the extra design effort and computational overhead associated with the more complex algorithms cannot be justified for simple process control applications. This reference also contains the results of applying random disturbances to control loops containing a wide selection of DDC algorithms. No deterministic controller appears to have performed well under these conditions. The conclusion is that a controller that is required predominantly to compensate for random disturbances should be designed as a stochastic controller.

Unbehauen's detailed conclusions can be summarised. Algorithms in Sections 5.1.1, 5.1.2 and 5.1.4 gave the best results in respect of response, sensitivity to parameter variations, design requirement, computer storage and computer time. Algorithms in Sections 5.1.6 and 5.1.7 proved to be too sensitive to changes in operating conditions and less efficient with step disturbances. All algorithms with low sampling rates were criticised for poor response to step disturbances – these included, of course, some of the most complex algorithms.

Further comparisons, also on a heat exchanger, have been carried out by Uronen and Yliniemi (1977). Their findings, which seem to agree with those of Unbehauen, can be summarised.

(*a*) For compensation against load disturbances, the PI algorithm has the best performance.

(*b*) For optimal following of set-point changes, the PID algorithm is best.

If T_1 is the dominant time-constant of the controlled system and T is the sampling interval, then a ratio $T/T_1 = 0.1$ gives a good performance, whereas a ratio $T/T_1 = 0.4$ gives a considerably degraded performance.

The dead-beat algorithm gives poor results with long settling times.

Dahlin's algorithm gives a slow response. It was concluded that the dead-beat algorithm and Dahlin's algorithm might be useful where a slow sampling rate ($T/T_1 \simeq 0.4$) has to be used. The following relative CPU times are quoted for the algorithms (PDP8/E computer).

> PID algorithm 74 ms
> Dead-beat algorithm 114 ms
> Dahlin algorithm 326 ms

A full-scale effluent neutralisation plant was controlled in an extended trial by Jacobs *et al.* (1980) and the discussion below is reproduced with permission.

The plant was originally controlled by a conventional PID controller which was judged inadequate, first because its parameters needed resetting frequently due to plant changes, and second because it could not cope successfully with the severe nonlinearity relating pH to concentration. These difficulties caused the controller to dispense more neutralising alkali than was strictly required. Three control schemes were evaluated: (S_1), PID with saturation prevention, (S_2), PID with saturation prevention, compensation of the known nonlinearities and some feedforward, (S_3), a self-tuning regulator with, in addition, the same saturation prevention, nonlinearity compensation and feedforward as for scheme (S_2).

In theory, scheme (S_3) should have been able to overcome the chief weaknesses of the original scheme by virtue of its ability to alter parameters and to make use of the stored curve describing the process nonlinearity. The following results were obtained in practice. To compare the operation of the three algorithms, two different performance criteria were defined as follows:

$$I_1 = \frac{1}{T} \int_0^T x(e)\,dt,$$

where

$$e = \text{deviation of pH from desired value}$$

$$x(e) = e, \quad e > 0$$

$$\quad\quad = 0, \quad e < 0$$

$$I_2 = \frac{1}{T} \int_0^T y(t)\,dt,$$

where $y(t)$ is the flow-rate of neutralising reagent. I_1 measures the quality of neutralisation while I_2 measures the cost of the reagent.

In trials lasting several months, the following results were obtained:

$$\frac{I_1(S_2)}{I_1(S_1)} = 0 \cdot 5,$$

$$\frac{I_1(S_3)}{I_1(S_1)} = 0 \cdot 49,$$

$$\frac{I_2(S_2)}{I_2(S_1)} = 0 \cdot 97,$$

$$\frac{I_2(S_3)}{I_2(S_1)} = 0 \cdot 83.$$

The paper reports that scheme S_3, self-tuning, was able to adapt to slow changes, but under unusual circumstances it was unreliable and then the process had to be switched back to PID control. The paper gives the impression that the S_3 algorithm needed to be supervised in case it suddenly drove off in the wrong direction.

5.4 Time and memory requirements of simple DDC algorithms

For a particular microprocessor the execution time for each type of operation or intruction is known. Hence, the total execution time for each control cycle can be estimated by summation of the individual instruction times. Let T_i be the basic instruction time for a particular processor and assume than an nth order linear controller with input u and output y is to be realised where

$$\frac{y(z)}{u(z)} = \frac{(a_0 + a_1 z^{-1} + a_2 z^{-2} + \ldots + a_n z^{-n})}{(b_0 + b_1 z^{-1} + b_2 z^{-2} + \ldots + b_n z^{-n})}.$$

Then the execution time T_c for each control cycle including overheads for input–output and other essentials is of the form:

$$T_c = (k_1 n + k_2)T_i,$$

in which, approximately, $k_1 = 20$ and $k_2 = 30$. (These numbers having been obtained by adding the operations involved in a typical program and rounding upwards.)

Thus, for a system based on an instruction time of $2\,\mu s$ and with a control algorithm of order 3 the estimate of cycle time using the above relation is $T_c = (60 + 30)2 = 180\,\mu s$. This calculation allows a rough estimate to be made of the maximum sampling rate that can be sustained using a microprocessor dedicated to the control loop.

Where algorithms are written in machine code it will be relatively easy to estimate the memory requirements. Where the programs are written in high-level language and a cross-compiler used to generate machine code for loading into PROM, memory-size estimation needs more detailed investigation.

5.5 Problems due to quantisation

The effects of quantisation should be examined to ensure that adequate control will be achieved under all conditions likely to be encountered. A first requirement is that a sufficient number of bits is used to give necessary accuracies. However, taking a digital PID algorithm as an example, other effects need to be considered:

(*a*) A quantisation step that is large in relation to the sampling interval may cause the output of the algorithm to be a series of high-amplitude spikes, perhaps driving plant actuators into saturation. (The incremental version of the algorithm, being the derivative version, can be expected to accentuate the effect. Noisy signals also aggravate the problem.)

As an illustration, consider a slowly rising ramp signal $e(t)$ input to an analogue/ digital converter for which $e = 2$ corresponds to 7 bits and sampling 1000 times/ second with $T_d = 4$ s in the PID algorithm (5.2). The derivative term gives out pulses of amplitude 62.5 volt and duration 1/1000 second, each time the ramp reaches a new quantisation level (Fig. 5.5). An analogue differentiator would have output a constant signal of 0·004 volts. The pulses may not have the same effect due to nonlinearities.

Even in the absence of nonlinearities, many processes will not tolerate kindly the jerky actions produced.

Notice two points:

(i)   Shortening the sampling interval accentuates the problem.

(ii) The incremental version of the PID algorithm (5.3) suffers doubly from the effect. The term $k(e_n - e_{n-1})$ behaves in the same pulsed manner that has just been discussed. The term $(e_n - 2e_{n-1} + e_{n-2})$ produces a pulse that is first positive going and then negative going at the point where the ramp reaches a new quantisation level.

(*b*) When a long integration time constant is to be set in the PID algorith, the effect of quantisation on the integral term needs to be examined.

If n-bit working is assumed and an integrator time constant $T_I = 2^p$ is set then, using a one second sampling interval, a threshold percentage of $(2^p/2^n) \times 100$ percent needs to be reached in the error signal, otherwise it will be ignored by the integrator.

This means that with 12-bit working and an integrator time constant of 2^{10} seconds (about 17 minutes), the integrator would ignore any errors of less than 25%!

Where an integrator time constant of an hour or so (say 2^{12} seconds) is needed, together with an accuracy requirement of better than 1%, this would imply by the same argument a necessary word length of at least 19 bits.

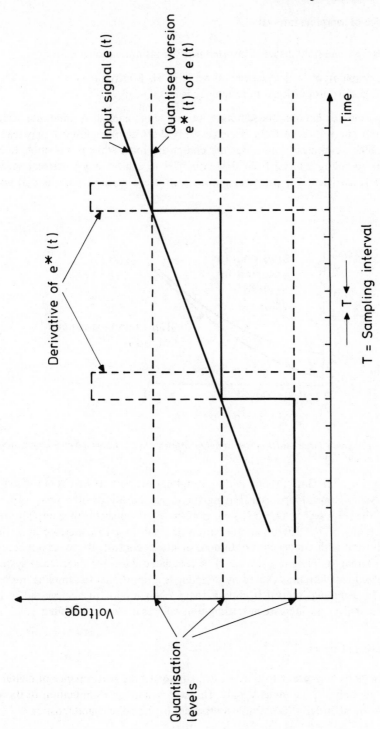

Fig. 5.5   *An illustration of the spiky output from the derivative term of a discrete-time PID algorithm*

5.6 Choice of sampling interval

When a sample-and-hold device is inserted into a continuous control loop:

(*a*) Intersample ripple at frequencies above $2\pi/T$ Hz is introduced.
(*b*) A phase shift proportional to frequency is introduced.

It is important to choose the sampling interval short enough to minimise system degradation due to these effects. An excessively short sampling interval may lead to noise problems and clearly increases the computing load. While it is possible to calculate the sampling interval from first principles to achieve any particular specification, it is sufficient for practical purposes to use a method such as that given below.

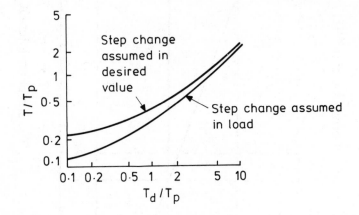

Fig. 5.6   *The graph (after Goff) allows sampling intervals to be chosen, given a simple process model*

Figure 5.6 after Goff (1966) allows sampling intervals to be chosen based on Ziegler–Nichols open-loop tests. Having found graphically or otherwise a process model of the form $ke^{-sT_d}/(1 + sT_p)$ the graph suggests values for the sampling interval T in terms of T_p. These sampling intervals result in a 15% increase in settling time compared with equivalent continuous analogue control. If 15% is not acceptable, a different sampling rate can easily be calculated since, for a particular system, the increase in settling time caused by sampling is proportional to sampling interval. Thus if the sampling interval is chosen to be one-tenth of that suggested in the figure, then only 1·5% increase in settling time will result due to sampling.

5.7 Filtering of signals

It is particularly important to prevent degradation of the performance of digital algorithms by noise in the input signals. This will require an examination of the signals to obtain an order of magnitude assessment of the noise characteristics.

Three broad categories of noise can be recognised. Uncontrollable process disturbance signals, measurement noise such as caused by liquid turbulence in flow measurement and electrical pick-up. The only useful information in a signal is that representing controllable disturbances. The choice of correct filter for noise removal requires an approximate knowledge of the noise spectrum, the process dynamics, and needs to take into account the sampling rate in the loop.

Empirical rules given in Goff (1966) allow the noise spectrum to be estimated from chart records as follows:

r.m.s. value of noise $\simeq$ (peak to peak value)/8,

time constant of noise characteristic $\simeq$ 0·8/(average number of zero crossings per unit time),

bandwidth of noise $\simeq$ (average number of crossings per unit time)/3·2.

Clearly, for difficult cases, a spectrum analyser should be brought into action, but the simple rules above at least allow a valuable initial orientation.

Chapter 6

'Standard' methods for the control of single-input/single-output processes

6.0 Introduction

This chapter is concerned with single-input single-output processes and the solution of their control problems by relatively straightforward methods. Much of the material can be found in classical control texts but supplementary material based on experience has been added.

Most of the chapter deals with process understanding, since control design for well-understood processes is relatively straightforward.

6.0.1 A note on non-minimum phase processes
Processes with right half plane zeros are called non-minimum phase processes. Such processes have a step response of the form shown in Fig. 6.1 and thus from a control point of view, they have some similarity with dead-time. Right half plane zeros cannot be cancelled by poles in the controller, since then the controller would be unstable.

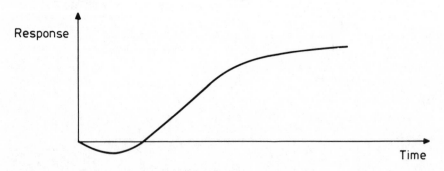

Fig. 6.1   *The step response of a typical non-minimum phase process*

Non-minimum phase processes occur rather rarely. One notable example is in electrical generation by hydroelectric turbines, where demand for additional power from a generator initially produces a drop in power until the water flow has reached a new steady state. A second example is given in Section 10.8.

Conventional frequency response techniques are not in general applicable unmodified to non-minimum phase processes and care should therefore be exercised in the analysis of these processes.

A continuous-time minimum phase system can become non-minimum phase on transformation to discrete-time form. The effect is sampling rate dependent and, as a rough guide, can be avoided by using a sampling period that is much shorter than any of the time constants in the system to be transformed.

6.1 Approaches to process modelling

The approaches can be classified loosely as follows:

(*a*) By fitting a simple model 'by eye' to a process step response curve.

(*b*) By fitting a transfer function or equivalent model numerically to agree with recorded process responses.

(*c*) By obtaining frequency response curves experimentally using a transfer function analyser and then fitting a transfer function numerically to the data.

(*d*) By obtaining a cross-correlation function from input—output data and then operating on this to obtain one of several possible model forms.

(*e*) By building up a model containing a core of equations representing the known mechanisms in the process and then determining unknown coefficients numerically (such a model is called a physically based model).

(*f*) By hypothesising relations and determining the magnitudes associated with them by regression.

(*g*) By defining stochastic dynamic relations and fitting unknown coefficients numerically.

Each method has its own particular application possibilities. Method (*a*) is used successfully, but without much publicity, by the process industries for setting up control loops. Method (*b*) is basically a computer-aided version of (*a*) and sharing the advantages, also offers the possibility of more accurate process description.

Frequency response methods (*c*) are highly successful in identifying electrical, mechanical or hydraulic processes, particularly where there are short-time constants and resonance effects. Other processes (thermal, chemical, etc.) have such long time constants that an excessive time would be required to carry out frequency response tests and long-term drifts during the testing period would tend to degrade the results. Cross-correlation techniques (*d*) are designed for difficult conditions, but in practice the results obtained are often disappointing. The physically-based model approach (*e*) is the most versatile and all-encompassing. In return for considerable extra modelling effort and an additional allowance of computer power, this approach offers consideration of process complexities, including non-linearities and operator intervention. A control engineer will often suggest advantageous modifications to the process, based on understanding gained through this type of model. Methods (f) are best suited to modelling steady-state situations. Methods (g) can be considered as an alternative to deterministic modelling. For instance, diffusion can be modelled on probabilistic assumptions about molecular velocities instead of using the deterministic partial differential equation of diffusion.

6.2 Obtaining a process model by correlation techniques

In many processes, it is forbidden to inject a test disturbance of significant amplitude and in the presence of inevitable process noise, meaningful identification is difficult if not impossible. In these circumstances, the idea of extracting a model statistically from a long record of normal process operation seems attractive. Usually, such an approach is not applicable to a real situation because of problems of statistical dependence between variables and poor signal to noise ratio.

The situation is much improved if a train of random disturbances is injected into the process and a correlation carried out between this random sequence and the perturbed output. The principle is that disturbances in the output due to causes other than the injected sequence will be independent and will, over a period, sum to zero.

Efficient portable commercial correlators are available to undertake this identification. Since true random number generation is very difficult to achieve, the correlators are used with pseudo random binary sequence (PRBS) noise generators. In use, the noise amplitude, frequency and sequence length have to be chosen and the cross-correlation function of the process is built up and displayed on the correlator screen. Slow drifts in the process tend to make the procedure inaccurate, so the length of the test sequence must be only long enough to ensure statistical meaningfulness.

The cross-correlation function of a linear process is identical with the impulse response. The cross-correlation function therefore gives valuable information directly – particularly on the general nature of the inter-relation between two variables, with a direct indication of any dead-time that may be present. For most purposes, however, the cross-correlation function will need to be converted into another form. There are three main possibilities:

(a) To select a suitable order of impulse response equation, and to fit its parameters numerically. (There are $n + 1$ parameters to determine where n is the order of the representation chosen.)
(b) To Fourier transform the cross-correlation function numerically into gain and phase estimates for the process, as a function of frequency.
(c) To set up a difference equation and to search for its coefficients numerically until its cross-correlation function agrees with that measured experimentally.

Large numbers of papers have been published in this general area and some good results have been reported in applications. Nevertheless in the 'intensely practical' field which we are constantly trying to keep in mind, it has to be said that the number of applications is disappointingly small. A valuable paper that uses real data from a plastics extruder is by Kochhar and Parnaby (1978). This paper illuminates some of the difficulties. In particular, the fitting of simple functions in approaches (*a*) and (*b*) is rarely adequate. However, approaches (*a*) and (*b*) give valuable preliminary information on dead-time and frequency response. Approach (*c*) seems to be unnecssarily circuitous, since the coefficients of the difference equation could be found directly from the input–output process data without first finding correlation functions.

6.3 Physically-based models

In this approach, the form of the process is described by a set of equations and the particular coefficients are determined numerically from process data. The approach therefore pre-supposes either that the process already exists to allow data to be obtained, or else that data from a similar process elsewhere can be used instead.

The framework of the model is a set of differential and algebraic equations representing the known theory of the process. For many processes, conservation equations (mass, energy, etc.) form the foundation of the basic set. Empirical relations based on observed results complete the model.

Coefficients in the equations are, if possible, found from the literature or from the results of specially devised tests.

The remaining unknown coefficients $a_1, \ldots, a_n$ are allocated feasible numerical values. Process input–output data are recorded over a wide range of operation. A hill-climbing routine is then used in the computer to manipulate the set of coefficients a_i until the model agrees as closely as possible with the process. The procedure is illustrated in Fig. 6.2. Experience shows that data accumulated from normal operating records are usually inadequate for the purpos of input to the hill-climbing routine: well-planned extensive data-logging trials are nearly always necessary in practice. The process does not need to be linear and it can have any number of input/output variables. The procedure really is robust and practicable. Practising engineers widely accept the accuracy and value of a properly constructed and tested physically-based model. The chief drawback of such models is the very considerable time required for their completion. (The author found some years ago that almost 80% of the time in control design projects was spent in modelling.) More insight into physical modelling can be gained from the case-history based reference of Nicholson (1980).

6.3.1 *Where to start in choosing the input–output variables for a physical model – an example*

An economic assessment of the process, as discussed in Chapter 2, will give some initial orientation. The next stages might be to obtain some feeling for the process variables and their inter-relation. To illustrate the idea, consider Fig. 6.3 which

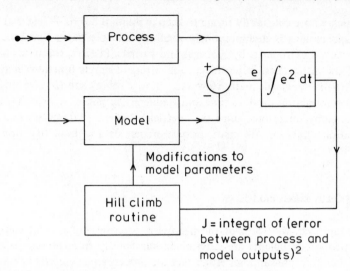

Fig. 6.2   *An iterative search for the unknown coefficients in a model*

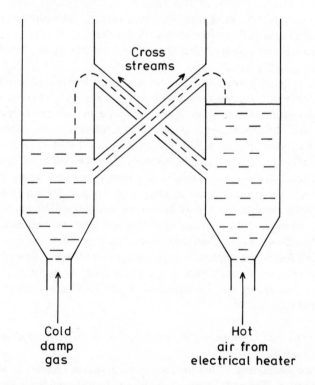

Fig. 6.3   *A double fluidised bed. Four blowers propel the two vertical and the two cross-streams of gas*

shows a double fluidised bed. In the left limb, cold damp gas is being dried through a bed of silica gel. In the right limb, used silica gel is being dried by a stream of hot air. The cross links allow continuous interchange of silica gel between the two limbs. Economics will have indicated what the control objectives might be, but where should one begin in representing the plant in a form that will be useful for subsequent controller design? Which variables should be considered as input, internal or output variables?

The mass of silica gel in each limb would appear to be of fundamental importance but only the levels are likely to be measurable and the level will depend not only on mass but on the extent of fluidisation. The humidity of the silica gel is also obviously of importance but how many parameters might be needed to describe the humidity state of all the silica gel? Again, there is an obvious measurement problem.

Turning to inputs it appears that suitable variables would be the voltages applied to the blowers and the voltage applied to the electrical heater. Notice, however, that another choice for input variables would be the, perhaps more fundamental, set consisting of four air pressures and one air temperature. This second set is related to the first set by relations involving bed permeability and heat transfer. Whichever set of input variables is chosen, somehow the mass flow rate and humidity of the gas to be dried will have to be included. This example has been chosen, because though it appears (and is) simple, it is by no means a trivial operation to make a start in characterising the process. There is, of course, no unique best way forward.

Once variables have been chosen, with the choice being influenced by measurement possibilities, some approximate relations can be established between these variables. Suppose a set of input variables $\{u_i\}$, $i = 1, \ldots, r$, internal variables $\{x_i\}$, $i = 1, \ldots, n$ and output variables $\{y_i\}$, $i = 1, \ldots, m$ have been chosen. Then the process can be visualised in linearised form, neglecting dynamics as

$$\begin{bmatrix} \Delta x_1 \\ \Delta x_2 \\ \cdot \\ \cdot \\ \cdot \\ \Delta x_n \end{bmatrix} = \begin{bmatrix} \dfrac{\partial x_i}{\partial u_j} \end{bmatrix} \begin{bmatrix} \Delta u_1 \\ \cdot \\ \cdot \\ \cdot \\ \cdot \\ \Delta u_r \end{bmatrix},$$

$$\begin{bmatrix} \Delta y_1 \\ \cdot \\ \cdot \\ \Delta y_m \end{bmatrix} = \begin{bmatrix} \dfrac{\partial y_i}{\partial x_j} \end{bmatrix} \begin{bmatrix} \Delta x_1 \\ \cdot \\ \cdot \\ \Delta x_n \end{bmatrix},$$

where $[\partial x_i/\partial u_j]$, $[\partial y_i/\partial x_j]$ indicate matrices, of appropriate order, of sensitivity coefficients.

The equations are separated because later, dynamics will be added to the first equation but, not usually, to the second.

The sensitivity coefficients can be estimated by special perturbation trials or from normal operating process records or by simulation if a set of steady-state process equations is available, as it often is.

6.4 Computer-assisted design of controllers

6.4.1 Design of controllers for linear single-input/single-output processes

The computer-aided design package described below was developed by the author. Many similar aids exist for linear system analysis and design. The package to be described can be understood by reference to Fig. 6.4.

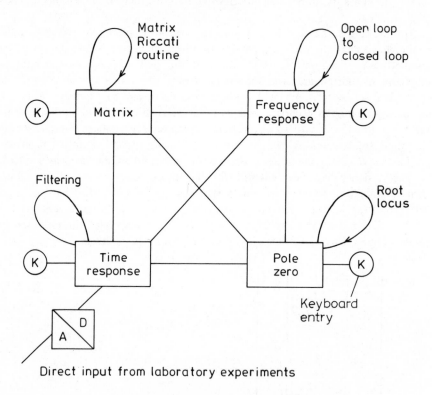

Fig. 6.4   *Schematic diagram illustrating the operation of the control design package*

The package is able to contain system representations in time response, frequency response, pole-zero, or matrix forms. System representations in any of these forms can be input via a keyboard while one direct signal input via an analogue to digital converter leads directly into the time response representation. Within each domain the operations shown in the table below are possible.

Domain	Operation
Time	Filtering
Frequency	Open-loop to closed-loop
Pole-zero	Root-locus and modified root-locus
Matrix	Design of minimum variance regulator

(The modified root-locus was developed for deciding on the correct gain for the integral term in a three-term controller. As such it keeps the proportional gain constant and varies the gain associated with the integrator, thus allowing the dynamic effects of different integrator gains to be appreciated.)

Transformations between the four domains are achieved very rapidly. Other features are: systems up to ninth order accepted, user interaction via a VDU, display interrogation via cross-wire cursor possible in all modes, hard copy of any display produced on demand.

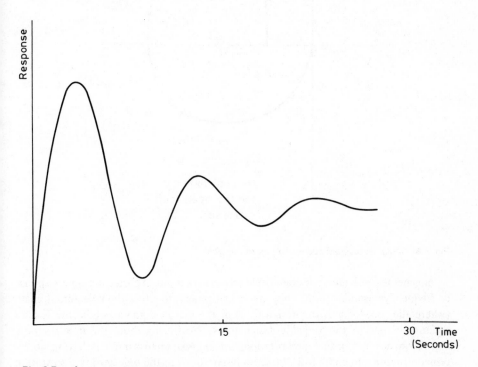

Fig. 6.5   *A process step response*

As an illustration of the use of the system, consider the simple case where a device whose step response is shown in Fig. 6.5 is to be controlled. The step response was fed directly into the time-domain via the analogue digital converter and then filtered to remove noise. (The filtering is performed by a very light moving average filter which, interactively, is applied repeatedly until the desired effect is obtained. In this case the filter was applied five times.) The frequency response corresponding to this time response is shown in Fig. 6.6 and root-locus diagram in Fig. 6.7. A matrix representation and an optimal feedback controller can also be calculated using the design package. Of course, the design process is iterative and the package is designed to allow repeated checks of progress and modifications to the design parameters.

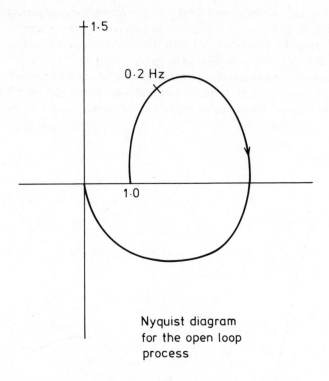

Fig. 6.6   *Nyquist diagram corresponding to Fig. 6.5*

In general, experimentally identified models are input in either the time response or frequency response form. They are transformed into either the pole-zero domain (where the root-locus is the design tool) or the matrix domain (where the Riccati equation is one of the available design tools). Notice that both the time response and frequency response representations are non-parametric models having no internal structure. In going from these representations to the pole-zero representation a system order needs to be decided upon, either manually or automatically. This is not a trivial decision. See Desai and Fairman (1971) for a description of one suitable method of deciding the order of a system. In moving further to the matrix form, internal structure has to be added in that state variables need to be chosen. This is relatively easy using canonical forms.

6.5 Understanding pole-zero plots

Despite the availability of computer packages to generate specific simulation results, it is still highly desirable to develop insight into the behaviour of a process through

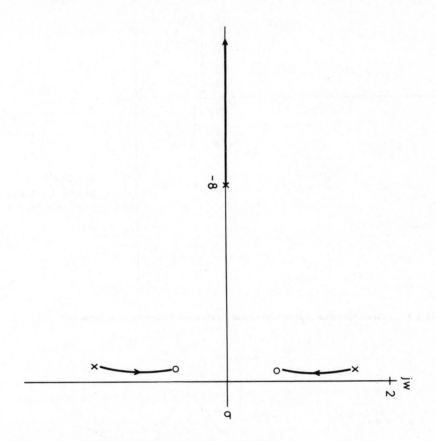

Fig. 6.7 *The root-locus diagram corresponding to Fig. 6.5*

representations such as the pole-zero diagram. Some useful initial guidelines are the following:

(*a*) Dominant poles are those nearest the origin of the pole-zero diagram.

(*b*) For many processes, the transient response is largely governed by the dominant complex poles. Quarter amplitude decay rate (accepted as a generally suitable setting) can be obtained by positioning dominant poles on the lines of slope 4·55 and −4·55 from the origin of the pole-zero diagram.

(*c*) In general, zeros in the transfer function affect the magnitude but not the nature of the process response.

(*d*) For a second-order process: undamped natural frequency ω_n, damped natural

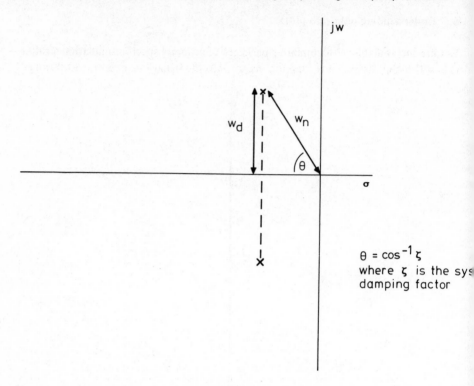

$$\theta = \cos^{-1}\zeta$$
where ζ is the sys
damping factor

Fig. 6.8   *Process response characteristics can be estimated from the pole locations*

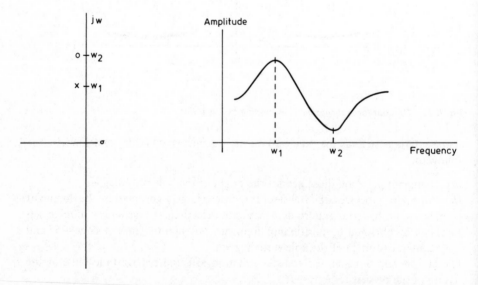

Fig. 6.9   *The process whose pole-zero diagram is shown has the frequency response curve of the form given*

frequency ω_d and damping factor ζ can be read off the pole-zero diagram as shown in Fig. 6.8.

(*e*) For low steady-state error, a pole of the controlled process should be near to the origin.

(*f*) Speed of response and degree of damping can be modified by moving dominant poles in the appropriate directions.

(*g*) A feeling for the frequency response of a process can be obtained as illustrated in Fig. 6.9.

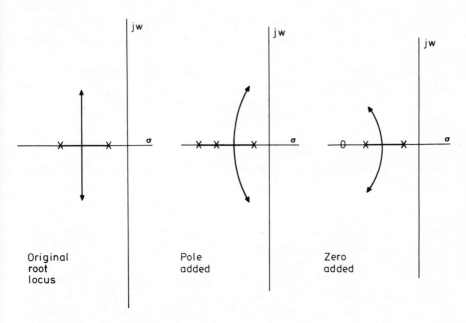

Fig. 6.10   *Qualitative effects of adding a pole or a zero*

Root-locus diagrams will normally be generated by a computer program. They constitute extremely useful design tools whenever they are applicable. In seeking to modify the shape of the root-locus by adding singularities, it is useful to remember the so-called proximity-effect (Dale-Harris, 1961). Roughly speaking, adding a pole to the left of a branch of the locus, deflects (repels) the locus to the right. Conversely, adding a zero to the left of a branch of the locus, deflects (attracts) the locus to the left (see Fig. 6.10). The rule can be a useful aid during interactive controller design using the root-locus technique.

6.6 Nonlinearities in control loops

Nonlinearities do not necessarily have an adverse effect on system behaviour — often quite the reverse. Witness how frequently designs evolved by nature contain nonlinearities. The chief reason for compensating for nonlinearity is that it allows one set of parameters to be fixed in a linear feedback controller to give a consistent performance over the whole process operating range.

6.6.1 Nonlinearities and stability

Saturation type nonlinearities and the smooth curve nonlinearities brought into a process by such physical devices as thermocouples rarely cause stability problems. It is the discontinuous small-signal nonlinearities such as backlash and hysteresis that do cause stability problems. Here classical methods, such as the describing function, can be used in an analysis of the effects — always provided that the process is not too complex. As always, simulation will be needed for processes too intractable to yield to analytic methods.

Nonlinear processes, by definition, have amplitude dependent gain. If the gain increases with amplitude, then disturbances may induce instability when the system is under closed-loop control. Nonlinearities that increase the loop gain at low amplitudes tend to cause continuous non-sinusoidal oscillations.

6.6.2 A simple general nonlinear process model

Smith and Groves (1977) have proposed a simple second-order nonlinear process model of the form

$$[a_1 + b_1 y(t)] y''(t) + [a_2 + b_2 y(t)] y'(t) + y(t) = [a_3 + b_3 y(t)] u(t).$$

Such a model assumes that the process can be represented by a steady-state gain, together with two time constants, where the gain and time constants vary linearly with $y(t)$. (Clearly this assumption will not in general be correct. However, the model goes some way towards accommodating process nonlinearities without introducing undue complexity.) The six unknown parameters are found by performing at least three perturbation tests of different magnitudes on the process and then using any differential equation solution routine in conjunction with a hill-climbing programme.

6.6.3 Nonlinear control strategies

For certain applications, a controller with a nonlinear gain characteristic can be applied with advantage. These applications include those where small deviations (for instance noise that cannot or should not be responded to) are to be ignored while large deviations are to elicit an extremely rapid response.

6.6.4 Velocity limiting and saturation in actuators

All positioning actuators are subject to a velocity limit and this represents a nonlinearity in the control loop. In a digital control loop the designer will usually arrange that the actuator travels at its maximum rate in performing each increment. An alternative arrangement must be made when a smoother actuator travel is required. This might be provided by arranging that the actuator completes its increment exactly at the end of each sampling period.

Hard saturation (for example, imposed by actuators that reach their physical limits) introduces severe nonlinearity into the control loop. Such saturation is, of course, a basic factor limiting system performance and as such it must not be neglected by making a linearising assumption.

6.6.5 Nonlinear control – a practical example

Given a process for which a fully validated mathematical model exists, it should be possible to determine by inspection where nonlinearities reside and to quantify their effects, at least approximately.

Consequently, we concentrate here on the alternative experimental approach to nonlinearities and illustrate the approach using a laboratory water-level control system as shown in Fig. 6.11.

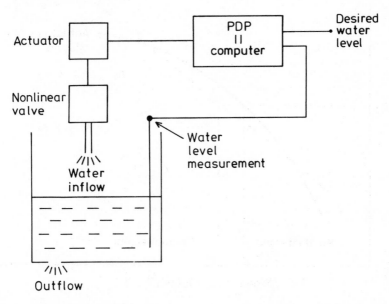

Fig. 6.11   *Water-level control system*

The valve has a curved characteristic of water flow against signal applied. Now it is an unnecessary sophistication for the computer to contain a full model of the water-level system. Let us assume that it in fact contains only a reasonably well set up three-term controller. The nonlinearity in the valve is taken account of as follows. Linearise the valve characteristic and for any particular controller output determine from the linear characteristic the flow q_0 that is being called for. Finally, adjust the actual controller output so that q_0 will result (see Fig. 6.12).

This strategy results in constant behaviour over the range of the nonlinearity and allows one satisfactory setting of the controller coefficients to span the range of operation.

This technique is clearly applicable to that class of processes having reasonably smooth nonlinearities. For saturation effects, simple logic will prevent the system being driven into the saturation region. Nonlinearities such as backlash or stick-slip motion tend to be difficult if not impossible to compensate properly by an algorithm (but see the method below using the describing function). Often the best solution is to penetrate the process mechanism and to seek modifications that will reduce directly the process nonlinearity. Here simulation will often be useful in forecasting the improvement in control that can be expected following different modifications to the process.

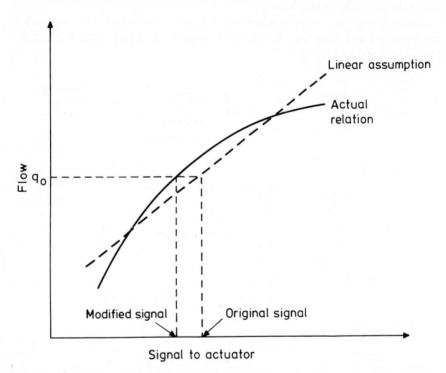

Fig. 6.12   *The nonlinear water valve characteristic and the method of compensating the non-linearity*

6.6.6 A control design technique based on the describing function [see Kallina (1981)]

The design of a feedback loop containing a hysteresis element is considered here. Some degree of hysteresis is present in most mechanical actuators — it can be the cause of many troublesome and puzzling oscillations.

Effects completely equivalent to hysteresis can occur where an actuator with dead-zone is part of a subsidiary closed-loop.

This troublesome and commonly arising phenomenon can be understood and compensated for by the use of describing function method (Grensted, 1962). This will allow estimation of the amplitude and frequency of oscillations and the estimation

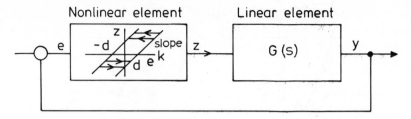

Fig. 6.13 *A control loop containing hysteresis — the hysteresis may arise from dead-zone in a minor loop.*

of the radius of the region of stability (if it exists). The method clearly shows the interacting effects of controller gain and actuator dead-zone on the oscillatory behaviour.

Figure 6.13 shows an element with hysteresis in a feedback loop in series with a linear element of transfer function $G(s)$. Recall that the hysteresis may arise from a minor loop containing a dead-zone. The transfer function G is taken to have the form shown in Fig. 6.14(*a*) — in the absence of the nonlinearity no instability problems occur in the loop.

Figure 6.14(*b*) shows the plot of the describing function for the nonlinear element as given by Grensted (1962). The plot is in terms of the dimensionless parameter a/d, where a is the amplitude of oscillation and $2d$ is the width of the hysteresis zone as given in Fig. 6.13.

The $G(j\omega)$ locus intercepts the describing function locus at two points, p_1 and p_2 [Fig. 6.14(*c*)]. The theory predicts that p_2 represents an unstable oscillation. Solutions in the neighbourhood of p_2 will either die away to equilibrium at the origin or they will grow until reaching p_1, which represents stable oscillation. Point p_2 therefore represents a stability limit.

By considering Fig. 6.14(*c*) a number of design modifications become obvious:

(*a*) Make d large. The point p_2 will then represent a large amplitude oscillation. Thus, the stability region will be increased. (This strategy leads to a conditional stability situation — it will need to be monitored with care.)
(*b*) Make d small and allow the system to operate at the point p_1, which will then represent a small amplitude oscillation that may be acceptable even though it occurs permanently.
(*c*) Attenuate the loop gain [Fig. 6.14(*d*)]. This strategy prevents oscillation but degrades the system response.
(*d*) Synthesise an amplitude dependent compensator to modify the describing function [Fig. 6.14(*e*)]. Teodorescu (1973) has suggested computer methods for such synthesis. Some degradation of small signal performance would be expected.
(*e*) Modify $G(j\omega)$ by classical frequency response compensation techniques [Fig. 6.14(*f*)].

The particular application will decide which of the techniques is preferable.

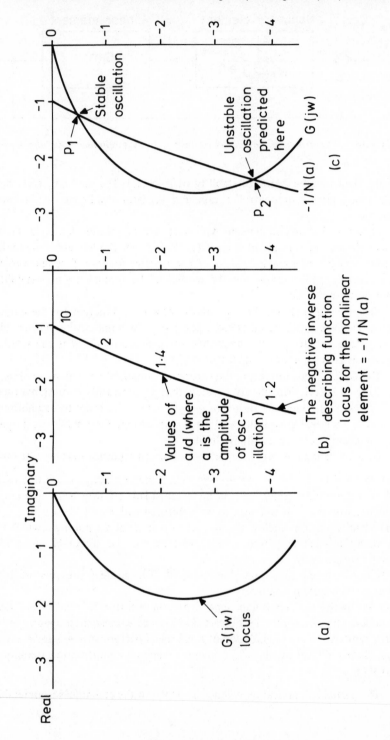

(a)

G(jw) locus

(b)

Values of a/d (where a is the amplitude of oscillation)

10
2
1·4
1·2

The negative inverse describing function locus for the nonlinear element = $-1/N(a)$

(c)

P_1

Stable oscillation

Unstable oscillation predicted here

P_2

$G(jw)$

$-1/N(a)$

Real

Imaginary

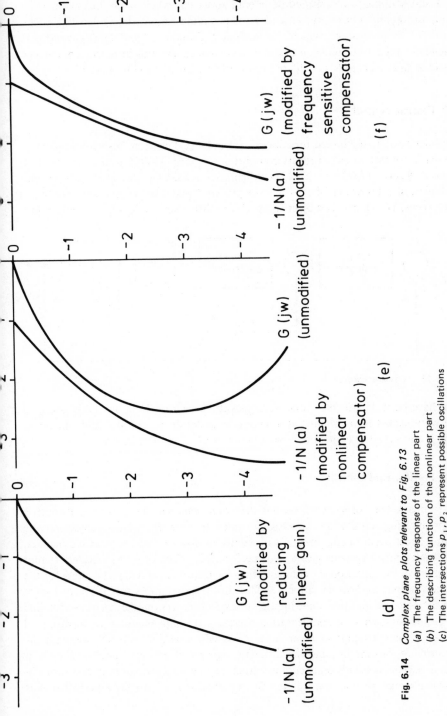

Fig. 6.14 *Complex plane plots relevant to Fig. 6.13*

(a) The frequency response of the linear part
(b) The describing function of the nonlinear part
(c) The intersections p_1, p_2 represent possible oscillations
(d) Oscillation prevented by reduction of loop gain
(e) Oscillation prevented by nonlinear compensator
(f) Oscillation prevented by frequency sensitive compensator

The describing function method is only applicable where, as in the example, decomposition into a linear dynamic part and a nonlinear nondynamic part is possible. Further, the premise on which the method is founded is that the dynamic part attenuates high frequencies more than low frequencies. For instance, a resonance effect in $G(s)$ is likely to invalidate the analysis by the method proposed.

6.7 Cascade control

Consider two closely-linked processes $G_1(s)$ and $G_2(s)$ in series. Instead of applying feedback control to each process conventionally, it is possible to nest the control loops as shown in Fig. 6.15. Such a configuration is called a cascade control system. Cascade control is best applied where the process $G_1(s)$ is fast-acting, compared with the process $G_2(s)$, and is subject to significant disturbances. Successful applications

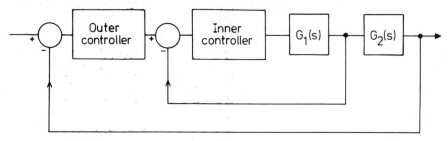

Fig. 6.15   *Cascade control*

that have been described include water-jacketted reactors. Here the temperature of the water-jacket is maintained at a desired value by the inner loop, while the outer loop is concerned to control the temperature of the reactor contents.

6.8 Feedforward control

In feedback control systems, output deviations can only be corrected through a loop involving actuator and process dynamics. In cases where large, rapidly varying disturbances occur in conjunction with relatively slow actuator or process dynamics, feedback control systems give a poor performance. Feedforward strategies attempt to improve the situation by compensating for disturbances without waiting for the process output to show deviations. Such compensation implies that the disturbances can be measured and that some implicit or explicit process model can be used on-line to decide the degree of compensation necessary. In implementing the feedforward strategy, some attempt must be made to synchronise the compensation with the disturbance, bearing in mind the different dynamic effects associated with each. If this is neglected, steady-state compensation may be achieved but large transient disturbances may go uncorrected. It will be unrealistic to aim for a perfect on-line model and the use of feedback control will nearly always be necessary to supplement feedforward control. Feedback control will also remain necessary to compensate for inevitable unmeasurable disturbances. Figure 6.16 illustrates the scheme of a combined feedforward/feedback system. Different methods of using the feedback signal are possible. Here the error is integrated through a low-gain feedback loop to compensate essentially for steady-state error in the on-line model. Such schemes can be very effective in practice.

6.8.1 Accomplishment of synchronism (refer to Fig. 6.16)

Define the two transfer functions

$$G_1(s) = y(s)/w(s), \quad G_2(s) = y(s)/u(s),$$

then the synchronism block will consist of a practicable approximation to the transfer function

$$G_3(s) = G_1(s)/G_2(s).$$

(This assumes that the process model takes negligible time to yield the signal v.)

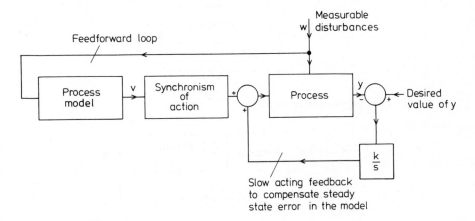

Fig. 6.16   *Feedforward/feeback control*

In summary, feedforward strategies may be used with advantage where:

(*a*) measurement or process dynamics make control difficult;
(*b*) an accurate process model exists that is feasible for rapid on-line use;
(*c*) the main control problem is to compensate for measurable input disturbances.

6.9 Methods for overcoming the effect of dead-time in control loops

By dead-time, we mean an effect representable by the transfer function e^{-sT} which delays a signal by time T but otherwise does not modify the signal. Such effects occur very frequently in measurements of output variables of industrial processes, where, necessarily, sensors must be mounted some way 'downstream' and the delay is often then referred to as a transport lag. Dead-time has a degrading effect on any control loop in which it is incorporated, as the following example illustrates.

6.9.1 Example

A process has the transfer function

$$G(s) = 1/s(s + 1)(s + 2).$$

If this process is put in closed-loop as shown in Fig. 6.17, a simple calculation shows that for stability $0 < k < 6$.

Now if the same process has a ten-second delay in series as shown in Fig. 6.18 it is found that for stability $0 < k < 0.228$. (This calculation was performed by finding the frequency ω^* at which the total phase shift over process plus delay is $-180°$ and then finding the value of k to keep the loop below unity at this frequency.) This example illustrates the very significant degradation of control that is brought about when a delay is unavoidably present in the control loop.

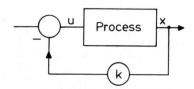

Fig. 6.17   *A control loop with no dead time*

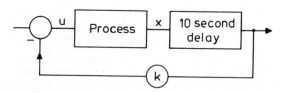

Fig. 6.18   *The same loop as in Fig. 6.17 except that a 10 s delay is now incorporated*

Figure 6.19 shows a method of overcoming the problem. The measurement model, which can be a simple linear model — valid only for perturbations, takes in ancilliary process variables and gives a near instantaneous estimate of the process output. This estimate is used for control purposes. The delayed measurement is used only to ensure that the process output has the correct mean level.

With this type of control, an unusual form of instability has been encountered. Ancilliary process deviations x_1, x_2 were input to a simple model to give an estimate $\hat{y}$ for the deviation y of the process output from its correct operating point. The model needed an estimate $\hat{c}$ of a positive process parameter c to yield its estimate

$$\hat{y} = x_1 + \hat{c}x_2.$$

If the situation arises where y and $\hat{y}$ have opposite signs, the control loop will be driven in the wrong direction! — this can occur if $\hat{c}$ is incorrectly specified as we now show. The process is such that x_1 and x_2 are of opposite signs.

We examine the expression

$$y\hat{y} = x_1^2 + c\hat{c}x_2 + (c + \hat{c})x_1x_2$$

and conclude that a value of $\hat{c}$ satisfying

$$\hat{c} > \frac{-x_1^2 - cx_1x_2}{cx_2^2 + x_1x_2}$$

will cause the process to be driven in the wrong sense. This can be shown to be a runaway situation since, however large the error becomes, the estimate $\hat{y}$ of the error has the wrong sign. Putting in typical values for the process as $c = 0.1$, $x_1 = -2$, $x_2 = 18$ yields $\hat{c} > 0.11$ for instability. The effect is asymmetrical — low values of c cause no stability problems. As always, once the problem is understood it is not difficult to build in appropriate safeguards.

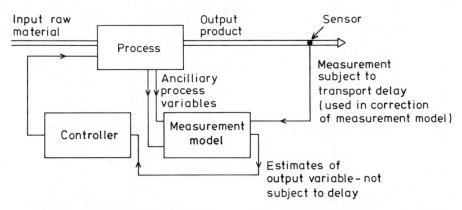

Fig. 6.19 *Compensation for process dead-time by using a measurement model operating on ancilliary process variables*

6.9.2 Compensation for process dead-time — the Smith predictor [Smith (1959) and Marshall (1979)]

Figure 6.20 shows the principle. If the process dead-time could be excluded from the control loop, all would be well, but this is in general not possible. However, if an on-line process model is available, an estimate of the undelayed output can be obtained from the model and this can be used in a control loop as shown in Fig. 6.20. The dotted lines show how provision is made for the use of a non-perfect model. In the example in the figure the error is treated in the simplest way possible, being added into the feedback signal from the model. A more sophisticated system uses the error to adjust the parameters of the model to match the plant. Attention may also have to be given to the effects of mismatch between the dead-times of

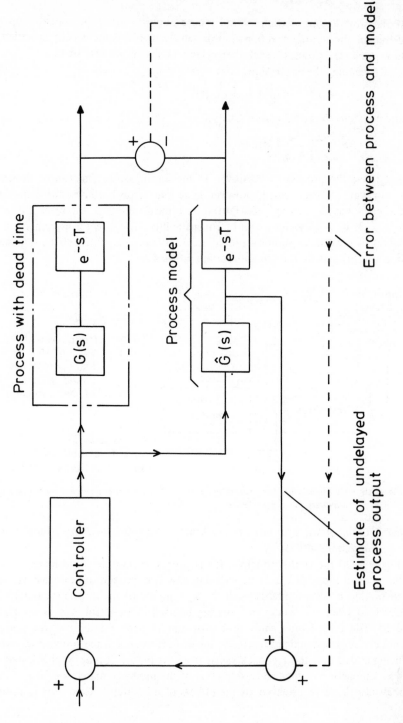

Fig. 6.20   *The Smith predictor*

process and model. The most difficult case is where the process dead-time varies rapidly over a wide range, since then it is extremely difficult to model accurately.

6.9.3 Modelling process dead-time

To allow analysis by root-locus technique dead-time e^{-sT} can be approximated by rational functions:

(a) $1 - sT$,

(b) $\dfrac{[1 - (sT/2)]}{[1 + (sT/2)]}$,

(c) $\dfrac{1}{[1 + (sT/n)]^n}$, $\quad n$ an integer to be fixed.

6.9.4 Comparison of controllers for processes with dead-time

Ross (1977) compares controllers for processes with dead-time having regard to costs, process variations, noise and ease of tuning as well as performance. His paper contains a potentially useful set of curves from which the parameters of a PI controller can be chosen, given a simple model of the process.

6.9.5 Extension to the multi-loop case

Donoghue (1977) discusses the problem of designing controllers for multivariable processes with time-delays. It will also be seen in Chapter 8 that the Inverse Nyquist design technique is able to encompass multi-variable processes with dead-time.

Chapter 7

Further techniques for controller design

7.0 Introduction

The three techniques described in this chapter are put forward for difficult situations where the methods of earlier chapters prove inadequate.

Self-tuning regulators should be considered where the process to be controlled alters its characteristics frequently in such a way that no one fixed controller can give good control. Reports indicate that self-tuning algorithms can be unrealiable on occasions. Some fail-safe mechanism needs to be built into the implementation such as an automatic switch to PID control, in the event of unusual behaviour being detected.

Synthesis of control loops of guaranteed stability is a topic of obvious interest if the techniques are applicable to practical situations. So far there are few reports of successful practical implementations but the technique has promise.

Control schemes including the predictive—iterative technique described in this chapter have been operating for several years. The technique is very widely applicable . The main point of decision is whether the relative complexity of the algorithm can be justified.

7.1 Self-tuning regulators for the control of single-input single-output processes [see Clarke (1979)]

7.1.1 Introduction
A self-tuning regulator, as usually applied, can be considered as an adaptive controller, changing its parameters to match changing plant characteristics. Although there are many other adaptive control techniques, the self-tuning regulator is described here because there have been a reasonable number of successful industrial implementations.

A self-tuning regulator comprises:

(*a*) A means of automatic process identification that yields values of parameters for insertion into a process model.

(*b*) An automatic control design procedure that, given the model from (*a*), produces and implements an on-line controller.

In use, the sequence: identify the process, choose the controller parameters, implement the controller, takes place within every sampling interval. The sampling interval must be chosen to allow these operations to take place, having regard to the types of signal and noise, the computing power available and the rate of adaptation thought necessary. A 16-bit microprocessor would be the usual device for realising the necessary algorithms.

7.1.2 *The automatic process identification*

The process model used in (*a*) is usually the discrete-time approximation of the *n*th order continuous linear model

$$\sum_{i=0}^{n} \alpha_i \frac{d^i y}{dt^i} = \sum_{i=1}^{n} \beta_i \frac{d^i u}{dt^i}, \tag{7.1}$$

(where we have, for brevity, used the convention that $d^0 y/dt^0 = y$).

The discrete-time approximation is expressible in the form

$$y_t = -\sum_{i=1}^{n} a_i y_{t-i} + \sum_{i=1}^{n} b_i u_{t-i}, \tag{7.2}$$

where the coefficients a_i, b_i depend on both the coefficients α_i, β_i and on the sampling interval.

If $2n$ input–output pairs (y, u) are sampled from time t onwards, then the two vectors y and u defined below are known to be

$$y = (y_t, y_{t+1}, \ldots, y_{t+2n-1})^T,$$

$$u = (u_t, u_{t+1}, \ldots, u_{t+2n-1})^T.$$

Define

$$\theta = (a_1, a_2, \ldots, a_n, b_1, b_2, \ldots, b_n)^T,$$

$$\Lambda = (-z^{-1}y, \ldots, -z^{-n}y, z^{-1}u, \ldots, z^{-n}u), \tag{7.3}$$

where z is the shift operator.

Then the $2n$ versions of Eq. (7.2) can be combined to yield

$$y = \Lambda \theta \tag{7.4}$$

And the vector θ of parameters can be obtained from the relation

$$\theta = \Lambda^{-1} y. \tag{7.5}$$

In practice this equation is not useful as an algorithm since the matrix inversion is easily ill-conditioned unless the (y, u) pairs are varying vigorously over the period of data collection.

If, however, p data pairs are recorded, where p is much greater than $2n$, Eq. (7.4) still applies. Multipylying both sides by Λ^T and reorganising yields

$$\boldsymbol{\theta} = (\Lambda^T \Lambda)^{-1} \Lambda^T y. \tag{7.6}$$

This equation will be recognised as a standard result in linear regression analysis and in fact Eq. (7.6) yields the best estimate of $\boldsymbol{\theta}$ in the least-squares sense.

7.1.3 Recursive least-squares algorithms

Equation (7.6) above has the (algorithmic) disadvantage that a large matrix has to be inverted on-line each time an estimate of $\boldsymbol{\theta}$ is required.

A recursive version can be developed as follows, provided that Eq. (7.6) is used to start the procedure.

Assume that the matrix Λ has been accumulated and that Eq. (7.6) has been used to obtain an initial estimate $\boldsymbol{\theta}_0$ of the parameter vector

$$\boldsymbol{\theta}_0 = (\Lambda^T \Lambda)^{-1} \Lambda^T y.$$

Assume that one further data pair (y_1, u_1) becomes available. The matrix Λ of Eq. (7.6) is expanded by the addition of one further row, say λ^T, and the new estimate, $\boldsymbol{\theta}_1$, of the $2n$ parameter vector is given by

$$\boldsymbol{\theta}_1 = \left(\begin{pmatrix} \Lambda \\ \lambda^T \end{pmatrix}^T \begin{pmatrix} \Lambda \\ \lambda^T \end{pmatrix} \right)^{-1} \begin{pmatrix} \Lambda \\ \lambda^T \end{pmatrix}^T \begin{pmatrix} y \\ y_1 \end{pmatrix}.$$

Using the following identity

$$(\Lambda_1^T \Lambda_1) = (\Lambda_0^T \Lambda_0)^{-1} - \frac{(\Lambda_0^T \Lambda_0)^{-1} \lambda \lambda^T (\Lambda_0^T \Lambda_0)^{-1}}{1 + \lambda^T (\Lambda_0^T \Lambda_0)^{-1} \lambda}$$

the desired recursive equation for $\boldsymbol{\theta}_1$ results:

$$\boldsymbol{\theta}_1 = \frac{\boldsymbol{\theta}_0 + (\Lambda^T \Lambda)^{-1} \lambda (y_1 - \lambda^T \boldsymbol{\theta}_0)}{1 + \lambda^T (\Lambda^T \Lambda)^{-1} \lambda}$$

while the matrix Λ is updated using the equation

$$\Lambda_1 = \frac{\Lambda_0 - \Lambda_0 \lambda \lambda^T \Lambda_0}{1 + \lambda^T \Lambda_0 \lambda},$$

where Λ_0 and Λ_1 are the old and updated matrices respectively.

7.1.4 Inclusion of dead-time in the model

Provided that dead-time in the process is known, constant and can be represented by k sampling intervals, then it can easily be included.

Equation (7.2) is modified to become

$$y_t = - \sum_{i=1}^{n} a_i y_{t-i} + \sum_{i=1}^{n} b_i u_{t-k-i}.$$

7.1.5 Nonlinear or time-varying processes

For a nonlinear process the same model is used but its parameters can be expected to vary as the operating point changes. Thus, both a nonlinear process and a time-varying process present a similar modelling problem in that the model parameters need to be able to change, perhaps rapidly, during the whole life of the algorithm and not during an initial tuning period. The rigorous theory of self-tuning regulators is based on assumptions of linear time invariant plants. The extension to nonlinear and time-varying processes, where model parameters have to be continuously tracked, has tended to be more empirical. Yet these nonlinear and time-varying applications are clearly the ones where self-tuning regulators have most to offer. It follows that the ability to handle the extension of the identification to cover these cases reliably will be a pre-requisite for practical applications. The various algorithms described for modelling time-varying processes all amount to using a data window so that old data have relatively little effect compared with current data. Case histories have demonstrated that the choice of data windows is both difficult and of critical importance for the success of the self-tuning algorithm.

7.1.6 Effects of noise
Process noise may cause bias in the parameter estimates. This problem can be overcome by incorporating a noise model in the algorithm.

7.1.7 The automatic control design procedure – illustration of the principles
Given a discrete-time model of the process to be controlled, as described in Sections 7.1.2–7.1.6, there are two basic approaches to controller design:

(*a*) To synthesise a controller that will minimise a given, usually quadratic, cost function. The simplest of these criteria occurs when the system output is to be kept constant in the presence of disturbances, the cost function is the variance of the process output and this strategy is referred to as minimum variance control.
(*b*) To synthesise a controller such that the poles of the resulting closed-loop system are assigned to given locations.

7.1.8 Minimum variance algorithm
Calculation of the control to minimise the output variance is a quadratic optimisation problem that can be solved by a number of methods once the process model is available. Considerable recent research activity has been devoted to development of efficient algorithms suitable for on-line use. Attention has been given to the non-minimum phase problem described below. Some recent methods integrate the modelling of the process and the calculation of the minimum variance control into a single algorithm that operates directly on process input—output data.

7.1.9 Problems with the minimum variance controller
Notice that the continuous process that is to be controlled is modelled in the discrete-time domain. This is important since many well-behaved processes have non-minimum phase discrete-time models, i.e. they have zeros outside the unit circle in the z-plane. A minimum variance strategy, in cancelling these process zeros by poles in the controller, tends to cause instability. This effect should be checked for. If necessary, the stability situation can be improved by extending the cost function to

include a cost on the control signal, as has been described by Clarke and Gawthrop (1979).

7.1.10 Pole placement algorithm

The poles of a known system can be moved to any desired locations (except in unusual circumstances when singularities prevent the operation) provided that the state vector is available to be fed back. In our application this means that the process output and all its derivatives up to the power of $n - 1$ must be available, and in fact they are estimated by difference approximations.

Assuming the state vector to be available, one method of pole placement requires the following steps.

The open-loop process model is transformed into the so-called second canonical form.

$$\dot{x} = \begin{bmatrix} 0 & 1 & 0 & 0 & \cdots & 0 \\ 0 & 0 & 1 & 0 & \cdots & 0 \\ \cdot & \cdot & \cdot & \cdot & \cdots & \cdot \\ \cdot & \cdot & \cdot & \cdot & \cdots & \cdot \\ \cdot & \cdot & \cdot & \cdot & \cdots & \cdot \\ 0 & 0 & 0 & 0 & \cdots & 1 \\ a_1 & a_2 & a_3 & a_4 & \cdots & a_n \end{bmatrix} x + \begin{bmatrix} 0 \\ 0 \\ \cdot \\ \cdot \\ \cdot \\ 0 \\ 1 \end{bmatrix} u.$$

The control u is then synthesised by

$$u = \begin{bmatrix} -\alpha_1 - a_1 \\ -\alpha_2 - a_2 \\ \cdots \\ \cdots \\ \cdots \\ -\alpha_n - a_n \end{bmatrix}^T x.$$

The closed-loop system then has the characteristic equation

$$\lambda^n + \alpha_n \lambda^{n-1} + \ldots + \alpha_1 = 0.$$

Clearly by choice of the coefficients α, the closed-loop eigenvalues can be located where desired.

7.2 Synthesis of systems of guaranteed stability by the second method of Lyapunov

A procedure was described in principle in Kalman and Bertram (1960) to synthesise a system so that Lyapunov's second stability theorem is automatically satisfied. The procedure can be illustrated by a simple example.

$$\dot{x}_1 = x_2,$$

$$\dot{x}_2 = -x_1 - x_2 + u,$$

$$u = -cx_1.$$

This is a second-order system connected in closed-loop. The coefficient c is to be chosen to give stable feedback control.

Define as a Lyapunov function $V = ax_1^2 + bx_2^2, a > 0, b > 0$. For stability we require that

$$\frac{dV}{dt} < 0 \quad \text{unless} \quad x_1 = x_2 = 0,$$

$$\begin{aligned} \frac{dV}{dt} &= 2ax_1 x_2 + 2bx_2(-x_1 - x_2 + u) \\ &= 2ax_1 x_2 + 2bx_2(-x_1 - x_2 - cx_1) \\ &= -2bx_2^2 + (2a - 2b - 2bc)(x_1 x_2). \end{aligned}$$

A sufficient condition for stability is that

$$a - b(1 + c) = 0,$$

$$a - b = bc,$$

or

$$c = \frac{a}{b} - 1.$$

According to the choice of coefficients a and b in the Lyapunov function, values for c anywhere in the range $(-1, \infty)$ will be found for the value of c ensuring stability. This example shows both the principle of the method and its possible weakness — that excessively conservative results might be obtained from an arbitrary choice of Lyapunov function. A search for the sharpest possible result requires a choice of 'best possible' Lyapunov function and there is often little guidance on how best to choose such a function.

Any finite positive pair of values (a, b) yield a positive definite V function whose time derivative is negative definite. Letting a and b range over all positive finite values in the last equation shows that, provided c is inside the interval $(-1, \infty)$, then the closed-loop system can be guaranteed asymptotically stable.

For simple systems, the Lyapunov method would not of course be used but the example illustrates the approach.

The method is applicable to any system to which Lyapunov itself is applicable so that, in particular, it is attractive for synthesising controllers for nonlinear processes.

Perhaps the most potentially useful application is in the synthesis of adaptive loops of guaranteed stability.

A practical point to watch is the following. Sampled data systems synthesised to have guaranteed stability have been found to take an inordinately long time to settle following a disturbance. Lyapunov guarantees only that the controlled system returns to its original position after a disturbance as time goes to infinity. The designer needs to intervene to ensure that the rate of convergence is satisfactory over the whole operational range of the system.

7.3 Predictive—iterative control using a fast process model [see Wood (1973)]

Any process for which a fast on-line model exists can be controlled by the technique shown in Fig. 7.1.

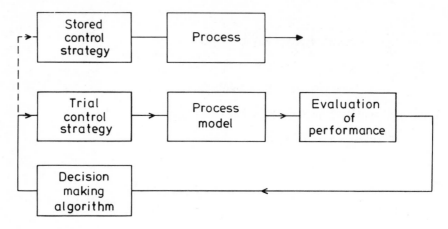

Fig. 7.1   *Predictive—iterative control strategy*

The process model, which should be several orders of magnitude faster than the process, receives trial control inputs, each defined over a simulated time interval T. The algorithm evaluates the consequent responses and chooses from the set of trial inputs the one that will be implemented on the actual process over the *next* time period T of process model time. Figure 7.2 below explains the inter-relation of model and process computations.

During the interval A of process time the control strategy is in use, chosen previously by the iterative use of the fast model in period 1. During the interval A, the model is being used in time period 2 and at the end of this period a control strategy will be found and implemented for the process to be controlled during interval B.

The advantages of the method are:

(*a*) No restrictions are imposed on the type of model provided that it satisfies the requirement of being suitable for fast on-line computation.

(*b*) A quite inaccurate model will still give good control. This is because at the beginning of each interval of process time T, the model starts off with the current initial conditions. The time T is chosen so that the prediction ability of the model is not overstretched.

(*c*) The difficult problem of control synthesis is avoided and no concessions need to be made to allow the synthesis problem to become tractable. Any criterion whatever can be used in the block marked 'evaluation of performance'. There is no need to be restricted to quadratic criteria.

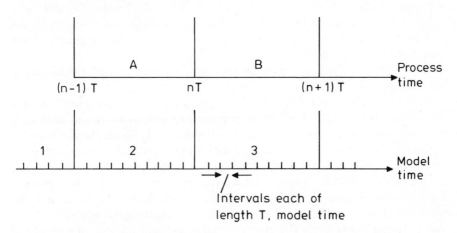

Fig. 7.2   *Predictive–iterative control strategy — relation between model time and process time*

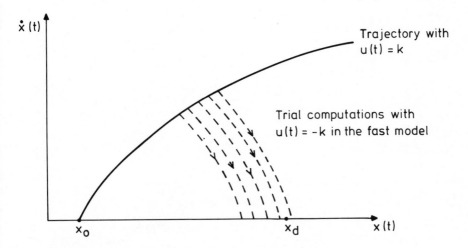

Fig. 7.3   *Predictive–iterative on–off control illustrated in the phase plane*

Individual schemes will have additional features according to the process control requirements. In particular, some means of updating the model so that it changes to match changes occurring in the process will often be required. Such additional features bring the system into the adaptive control area. Despite such additions, the basic principle of the predictive–iterative control remains as above.

We complete the description by giving an example that illustrates the versatility and basic simplicity of the technique. Suppose that a mechanical second-order system is to be brought from an initial position x_0 to a given final position x_d in minimum time by the application of a bang-bang control $u(t)$ that takes on either the value $u(t) = k$ or $u(t) = -k$. (Provided that the mechanical system has real eigenvalues then the minimum time control will contain a single switching between the two control possibilities — it is the task of the control strategy to determine and implement this change-over.)

The behaviour of the system under predictive—iterative control is illustrated in the phase plane in Fig. 7.3.

In Fig. 7.3, it is assumed for the sake of simplicity that the first part of the control strategy where $u(t) = k$ is already implemented and the control problem is to determine the change-over time to the control $u(t) = -k$. During the rise along the trajectory where $u(t) = k$, trial computations are made using the control $u(t) = -k$ on the fast model. These are repeated as often as the speed of computation will allow and the resulting computed trajectories are sketched in the phase-plane as dotted lines. To complete the system a logical block detects the arrival of a trial trajectory close to x_d. This block then switches the actual control to $u(t) = -k$ and this completes the control. This type of strategy has achieved successful control of a fourth-order mechanical system with only a very crude fast model and bang-bang control, much as outlined above except for non-fundamental complications incurred by the higher order.

Chapter 8

Methods for the analysis and design of multi-loop processes

8.0 Introduction

Interconnected processes interact dynamically in a way that needs to be understood from a physical point of view. Consider the case of three interconnected water tanks (Fig. 8.1). For the purposes of this analysis we can assume linear relations between head and outflow or crossflow. We obtain three linear differential equations which can be diagonalised. These equations show the fundamental modes of dynamic behaviour. As expected, there are three modes in this case, corresponding to the three eigenvalues $\lambda_1, \lambda_2, \lambda_3$:

Mode 1, in which the levels of all three vessels rise and fall together. The time constant of this mode is $1/\lambda_1$ — where λ_1 is the smallest eigenvalue — this is a much longer time constant than that of any of the individual vessels.

Mode 2, in which the level in the central vessel remains fixed while the levels of the two outer vessels oscillate in anti-phase.

Mode 3, in which the level in the central vessel moves in one direction, while levels in the two outer vessels move together in the opposite direction.

Note that instead of having three similar time constants, there is one large time constant, mathematically representing a single capacitance, while the two other time constants indicate the presence of interactive modes in the configuration.

In developing a computer control scheme for a large process with many loops, the main problem confronting the system designer is not that of control algorithms. More important in general are the problems of establishing a reliable data-base to include a wide variety of information, not all of it emanating from sensors, an efficient means of man/machine interaction and the incorporation of reliability/self-diagnosis facilities. Heavy interaction is likely to interconnect groups of up to five or six individual loops and algorithm design for these groups is fairly easily achieved using some of the methods to be outlined in this chapter.

A physically-based simulation in which the model parameters correspond with actual process parameters is a great help in deciding on the best configuration for the control system and investigating details that defy analytic solution.

The methods put forward here are first, empirical methods primarily suited to steady-state interaction assessment. Next, steady-state decoupling by diagonalisation and the extension to dynamic decoupling are treated.

Controller design through eigenvalue modification finds application in electro-mechanical processes where accurately known equations are available.

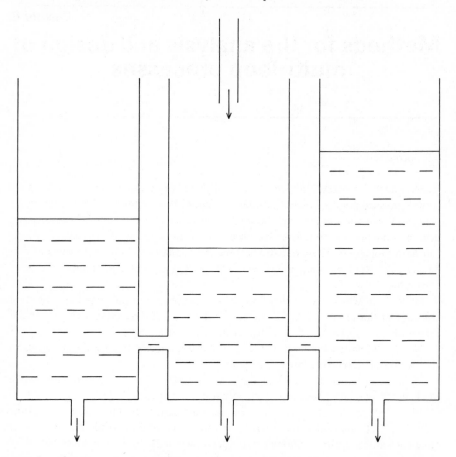

Fig. 8.1   *Interconnected water tanks*

Finally, Richalet's method of design, which has been proved valuable in realistic industrial situations, is described.

Two useful references on multivariable design are:

(*a*) Fisher and Seborg (1976) who applied different techniques to the control of a pilot plant evaporator;
(*b*) Sayers and Moore (1977) who presented a simplified account of multivariable controller design using a first-order discrete-time matrix decoupling method, followed by the use of PID controllers in individual loops. At least on a laboratory-scale plant, their approach appears to have produced good results.

8.1 The relative gain — a measure of steady-state interaction

For a multivariable interacting process, a useful tool in the early stages of investigation is the concept of *relative gain*, introduced by Bristol (1966). Let a process in the steady-state be represented by the equations

$$x_i = f_i(x_i, \ldots, x_n, u_1, \ldots, u_n), \quad i = 1, \ldots, n.$$

The open-loop steady-state sensitivity of state variable x_j to control input u_k is defined to be $(\partial x_j/\partial u_k)_0$.

The closed-loop steady-state sensitivity of state variable x_j to control input u_k is found by assuming that all other values of $x_i, i = 1, \ldots, n, i \neq j$, are held constant by closed-loop operations. Substituting to take this into account yields the equation of closed-loop operation. Differentiating the closed-loop equation yields the closed-loop steady-state sensitivity of state-variable x_j to control input u_k which is defined to be $(\partial x_j/\partial u_k)_c$.

The relative gain for the relation between u_k and x_j is then defined to be

$$\alpha_{jk} = \left[\frac{\partial x_j}{\partial u_k}\right]_0 \Bigg/ \left[\frac{\partial x_j}{\partial u_k}\right]_c.$$

The system described above has n^2 relative gains which can be arranged in a square matrix having the property that each column and each row sum to unity. Inspection of the matrix yields valuable information on the nature of the control problem and on the degree of interaction between variables. When $\alpha_{jk} = 1$, there is no interaction between this particular control mechanism and any other. If all the values are positive, then clearly they all satisfy the relation $0 < \alpha_{ij} < 1$ and under this condition the system can be expected to be stable. Control loops will normally be chosen to operate between those variables linked by the largest values of α.

Negative values of α in general indicate a more difficult control problem with possibilities of instability for at least some of the possible control configurations.

8.1.1 Illustrative example

A process with state variables x_1 and x_2 has input controls u_1 and u_2 and is described in the steady state by the equations

$$x_1 = u_1 + \frac{u_2}{2},$$

$$x_2 = -0.075u_1 + 0.1u_2.$$

The closed-loop equations are, for x_1 assuming that x_2 is held constant,

$$x_1 = u_1 + \frac{10x_2 + 0.75u_1}{2} = \tfrac{4}{3}u_2 - \tfrac{40}{3}x_2 + \tfrac{1}{2}u_2,$$

and for x_2 assuming that x_1 is held constant,

$$x_2 = -0.075u_1 + 0.1(2x_1 - 2u_1) = -0.075\left(x_1 - \frac{u_2}{2}\right) + 0.1u_2,$$

$$(\partial x_1 / \partial u_1)_0 = 1, \quad (\partial x_1 / \partial u_1)_c = 1.375, \quad \alpha_{ij} = 0.727,$$

and the relative gain matrix is determined from the property that rows and columns sum to unity.

The matrix of relative gains is found to be

$$\begin{bmatrix} 0.727 & 0.272 \\ 0.272 & 0.727 \end{bmatrix}.$$

In this example, control u_1 can control variable x_1 and control u_2 can control variable x_2 without stability problems being expected to occur. The interaction can be seen to degrade the control performance. Notice that dynamic effects were not considered at all.

Considering a different set of process equations

$$x_1 = u_1 + 2u_2,$$

$$x_2 = u_1 + u_2,$$

yields the matrix of relative gains.

$$\begin{bmatrix} -1 & 2 \\ 2 & -1 \end{bmatrix}$$

The minus signs indicate a change of polarity in the appropriate sensitivity coefficient between the open- and closed-loop conditions. More care is needed with this type of control problem than in the case where all the α-coefficients are positive.

The decision may be made to design a decoupling controller to obtain a more favourable control situation. However, analysis shows that processes having relative gains outside the $(0, 1)$ range and under decoupled control are excessively sensitive to small changes in parameters and may, in a practical situation, easily be brought into an unstable region by such small parameter changes [see Shinskey (1979)].

8.2 Simple methods for designing controllers when interaction is present

Where two control loops interact dynamically in a potentially destabilising manner, then if the speeds of response of the two loops are quite dissimilar, it is known from experience that the system can be stabilised by reducing the gain in the slower loop. Careful choise of the variables to be controlled, in conjunction with a consideration of the relative gain matrix, is often sufficient to produce a configuration where the interaction is low enough to allow good control to be achieved on an individual loop by loop basis. Where interaction between loops needs to be removed, this can be achieved for problems of low dimension (2 or 3 loops) by inspection. For instance, Shinskey (1979) argues that, to control the gas flow and pressure in the example of Fig. 8.2 requires a flow controller to move both valves in the same direction, a pressure controller to move them in opposite directions and some dynamic elements to match the transient behaviour. This example illustrates the general approach to multivariable control system design. Each output deviation causes a change in all actuators to compensate the steady-state interaction. Dynamic control elements need to be inserted to ensure that compensatory actions are in phase with the disturbances they are intended to counteract.

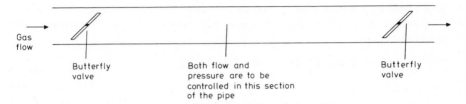

Fig. 8.2   *A simple multivariable control problem*

8.3 Static decoupling

8.3.1 Decoupling algorithms – motivation
Consider a linear system described by the equations

$$\dot{x}(t) = Ax(t) + Bu(t),$$

$$y(t) = Cx(t).$$

Let an error vector e be defined by

$$e(t) = v(t) - y(t),$$

where $v(t)$ is the desired (vector) value of $y(t)$. If now

$$u(t) = Ke(t),$$

where K is a diagonal matrix, then the whole forms a multivariable feedback control system. (We have assumed that all vectors are n-dimensional for simplicity.) We can imagine that the process operator intervenes to control the process through the desired value vector v. However, because of cross-coupling in the A, B and C matrices, the process response as shown in $y(t)$ is not related in a simple way to changes made in the v vector and the operator may have difficulties in achieving the results he requires.

8.3.2 Decoupling algorithm – the approach
Consider again the system as above.

$$\dot{x}(t) = Ax(t) + Bu(t),$$

$$y(t) = Cx(t),$$

$$e(t) \ = \ v(t) - y(t),$$

but now define

$$u(t) \ = \ De(t)$$

The matrix D is to be chosen so that *in the steady-state* a change in the element v_i in the v vector affects one and only element y_i in the y vector. We call this *static decoupling*. It is relatively easy to achieve and has considerable benefits where plants with significant cross-coupling are subject to frequent operator interaction.

Consider the whole system as just described to be represented by a matrix transfer function $T(s)$, i.e.

$$y(s) \ = \ T(s)v(s).$$

Then clearly if $T(s)$ were diagonal, every element $v_i(s)$ in $v(s)$ would be connected with only its corresponding element $y_i(s)$ in $y(s)$ and complete decoupling would have been achieved. Such decoupling may sometimes be justified but here we restrict ourselves to the steady-state. Suppose that each element $v_i(s)$ in $v(s)$ is a step function of amplitude α_i, then the corresponding solution for $y(s)$ is

$$y(s) \ = \ T(s) \ \frac{1}{s} \begin{bmatrix} \alpha_1 \\ \cdot \\ \cdot \\ \cdot \\ \alpha_n \end{bmatrix} .$$

Then applying the final value theorem

$$\underset{t \to \infty}{y(t)} \ = \ \underset{s \to 0}{T(s)} \begin{bmatrix} \alpha_1 \\ \cdot \\ \cdot \\ \cdot \\ \alpha_n \end{bmatrix}$$

and provided that $T(s)$ is diagonal then the system will be statically decoupled in that, in the steady-state, each element in the v vector will only affect the corresponding element in the y vector.

Now

$$T(s) \ = \ C(sI - A)^{-1} BD [C(sI - A)^{-1} BD + I]^{-1},$$

so that if the matrix D is chosen to satisfy

$$D \ = \ (CA^{-1}B)^{-1} \begin{bmatrix} k_1 & & \\ & \cdot & \\ & & \cdot \\ & & & k_n \end{bmatrix},$$

for some finite coefficients k_i, $i = 1, \ldots, n$, then the resulting system will be statically decoupled.

The values of the k_i still need to be chosen to give good dynamic response, so that the final stages are to choose these k values so that the closed-loop poles are satisfactorily located. The approach of this section will be found to be contained within the next section, which deals with both static and dynamic decoupling.

8.4 Dynamic decoupling – the Rosenbrock inverse Nyquist array techniques

The inverse Nyquist array technique operates in the frequency domain and has shown itself to be well-suited to the practical design of controllers for multivariable interacting processes. Well-established interactive CAD packages exist through which engineers can iterate towards controller designs that satisfy interaction criteria and engineering constraints simultaneously. Full details of the approach described below can be found in Rosenbrock (1969, 1974).

The process to be controlled is described by an $m \times m$ matrix $G(s)$ of rational transfer functions (but note that non-square matrices also fit the formulation, by choice of K and that dead-times, modelled by e^{sT}, can be incorporated). $K(s)$ is a possibly dynamic controller and F is a constant diagonal feedback matrix (see Fig. 8.3).

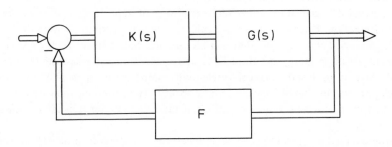

Fig. 8.3 *Multivariable control design – G is the process to be controlled, K is the precompensator, F is the (diagonal) feedback matrix*

The closed-loop transfer matrix $H(s)$ of the system is

$$H(s) = [I + G(s)K(s)F]^{-1}G(s)K(s)$$

and provided that $G(s)K(s)$ is invertible

$$H(s)^{-1} = [G(s)K(s)]^{-1}[I + G(s)K(s)F] = K(s)^{-1}G(s)^{-1} + F.$$

Let $Q(s) = G(s)K(s)$, then $H(s)^{-1} = Q(s)^{-1} + F$.

The design method is to determine $K(s)$ such that $Q(s)^{-1}$ is diagonally dominant. (An mth order matrix A is diagonally dominant if

$$|a_{ii}| > \sum_{\substack{j=1 \\ j \neq i}}^{m} |a_{ij}|, \quad i = 1, \ldots, m.$$

Where the elements of the matrix in question are functions of the complex variable s, diagonal dominance implies satisfaction of the above criterion for all values of s on the infinite right half plane.)

Denote the individual elements of the matrix $Q(s)^{-1}$ by q_{ij}^{-1}.

For interactive design, diagonal dominance is checked for by a simple geometric method. First a locus of $q_{ii}^{-1}(\omega)$ is plotted in the complex plane. This is the usual inverse Nyquist diagram and on this locus are superimposed circles of radius

$$r(\omega_k) = \sum_{\substack{j=1 \\ j \neq i}}^{m} |q_{ij}^{-1}(\omega_k)|$$

for user selected frequencies ω_k. These are called *Gershgorin circles*. If all such circles exclude the origin of the complex plane and the $(-1, 0)$ point and this is true for all the plots of $q_{ii}^{-1}(\omega)$ for $i = 1, \ldots, m$, then the open-loop system is diagonally dominant. The stability of the complete system can be ensured provided that each individual loop satisfies classical, single-loop, stability criteria.

In the absence of any other guidelines, a useful starting point towards achieving diagonal dominance is to achieve zero frequency diagonal dominance. This can be achieved easily by setting $K^{-1} = G(s)_{s \to 0}$, provided that det $[G(s)] \neq 0$. (det is a standard abbreviation for determinant.)

Diagonal dominance over the frequency range of importance is then achieved by user interaction. Once this stage has been reached, the maximum values of the feedback gains f_i, $i = 1, \ldots, m$, that can be used are found from the inverse Nyquist diagrams. If these gains f_i are thought to be too low for good response, the next and final stage is to insert classical single-loop compensators such as phase advance blocks in the invididual loops where necessary. The proposed final design will be checked by performing time simulations of the system under typical operating conditions.

Where a digital controller is to be designed for a continuous process with transfer matrix $G(s)$, the same basic approach is used but the following modifications are made.

$G(s)$ is transformed to discrete form by the substitution $z = e^{sT}$ where T is the chosen sampling interval. By making the further substitution $\omega = (1 + z)/(1 - z)$ where $\omega = u + jv$, frequency response methods are applicable to $Q(jv)$, v now being the analogue of frequency ω in the continuous case discussed earlier (see Munro, 1975).

8.4.1 Simple multivariable design example

The physical background and context for this example is given in Section 10.10. The open-loop process has the transfer matrix

$$G(s) = \begin{bmatrix} \dfrac{6}{s(1 + 0 \cdot 4s)} & \dfrac{-60}{s(1 + 0 \cdot 05s)} \\[4mm] \dfrac{-3}{s(1 + 0 \cdot 4s)} & \dfrac{160}{s(1 + 0 \cdot 05s)} \end{bmatrix}.$$

8.4.2 Analysis by the relative gain method

Let the two states be x_1, x_2 and the two control inputs be u_1, u_2. Then, differentiating the process equations with respect to u_1 and u_2 yields the open-loop sensitivity coefficients

$$\frac{\partial \dot{x}_1}{\partial u_1} = 6, \quad \frac{\partial \dot{x}_1}{\partial u_2} = -60, \quad \frac{\partial \dot{x}_2}{\partial u_1} = -3, \quad \frac{\partial \dot{x}_2}{\partial u_2} = 160.$$

Next, assume that ideal closed-loop operation obtains. After making the appropriate substitutions as described in Section 8.1 and calculating the closed-loop sensitivity coefficients we obtain

$$\frac{\partial \dot{x}_1}{\partial u_1} = 4{\cdot}875, \quad \frac{\partial \dot{x}_1}{\partial u_2} = 260, \quad \frac{\partial \dot{x}_2}{\partial u_1} = 13, \quad \frac{\partial \dot{x}_2}{\partial u_2} = 130.$$

Thus we obtain the relative gain matrix

$$\begin{bmatrix} 1{\cdot}231 & -0{\cdot}231 \\ -0{\cdot}231 & 1{\cdot}231 \end{bmatrix}.$$

This matrix indicates:

(*a*) The diagonal terms dominate.
(*b*) There is significant interaction indicated by the off-diagonal terms.
(*c*) The fact that some terms are greater than unity whereas others are negative, indicates that the interaction could cause stability problems.

8.4.3 Analysis of the example using the inverse Nyquist array

If we try to choose, as in Section 8.4, a pre-compensator K to achieve steady-state decoupling using the relation

$$K^{-1} = \underset{s \to 0}{G(s)}$$

division by zero results. However we can, with some physical justification, use instead the relation

$$K^{-1} = \underset{s \to 0}{sG(s)}.$$

This results in

$$K = \begin{bmatrix} 6 & -60 \\ -3 & 160 \end{bmatrix}^{-1} = \frac{1}{780} \begin{bmatrix} 160 & 60 \\ 3 & 6 \end{bmatrix}.$$

This is the controller to achieve steady-state decoupling as discussed in Section 8.3.

We take this as our starting point for dynamic decoupler design and after some computer-aided iterations we achieve diagonal dominance as required.

Space does not allow all the frequency response iterations to be reproduced.

We reproduce only Fig. 8.4, showing the inverse Nyquist plots with superimposed Gershgorin circles for the process with the controller that gives steady-state decoupling.

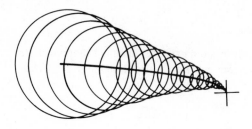

Fig. 8.4   *Inverse Nyquist array with Gershgorin circles for the strip thickness/flatness control problem with steady-state decoupling controller*

8.4.4 Time domain responses – before and after compensation
The figures show closed-loop step responses for this example with:

(*a*) Steady-state decoupling compensator (Fig. 8.5).
(*b*) Compensation to achieve diagonal dominance (Fig. 8.6).

8.5 Controller design through eigenvalue modification

If a controllable linear process, describable by state space equations, has eigenvalues $\lambda_1, \ldots, \lambda_n$ and if all the process states are available (i.e. can be either measured or estimated) then by a theorem due to Wonham (1967) any subset $\{\lambda_i\}, i = 1, \ldots, r$, $r \leqslant n$, of the eigenvalues can be changed to new values $\lambda_1, \ldots, \lambda_r$ specified by the designer by the use of appropriate linear feedback loops. Such a procedure can be useful for example in modifying the natural responses of a device such as an aircraft to produce desirable handling qualities.

However, the approach in which a controller is synthesised to achieve desired closed-loop poles has found little favour in industrial applications. Some of the reaons for this disfavour are:

(*a*) The state vector must be available to be fed back.
(*b*) A good model of the process must be available.
(*c*) The resulting system may be unacceptably sensitive to process parameter changes.

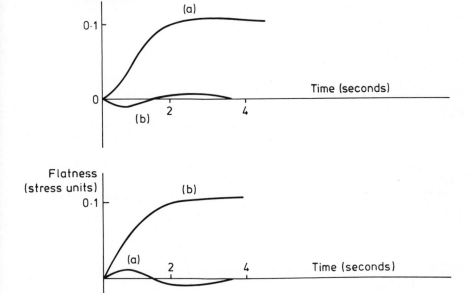

Fig. 8.5 *The performance of the rolling process with the controller designed to give steady-state decoupling*
(*a*) response to a step demand in thickness
(*b*) response to a step demand in flatness

A general algorithm for eigenvalue modification (also called a pole-placement algorithm) involves a number of algebraic transformations. However the principle is very simple. Given the desired closed-loop eigenvalues, form the desired closed-loop characteristic equation. Then choose a feedback matrix so that the desired characteristic equation results.

For a second-order process, the feedback matrix can be determined by inspection as the example below illustrates. (For higher order processes a systematic method is required – see Wiberg (1971), solved problem 8.2.)

8.5.1 Eigenvalue modification – example
The process is described by the equations

$$\dot{x} = \begin{bmatrix} 0 & 1 \\ -3 & -4 \end{bmatrix} x + \begin{bmatrix} 1 & 1 \\ 0 & 1 \end{bmatrix} u = Ax + Bu. \qquad (8.1)$$

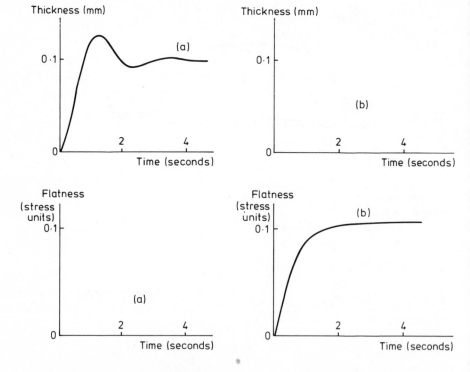

Fig. 8.6   *The performance of the rolling process with the dynamically designed controller*
(*a*) response to a step demand in thickness
(*b*) response to a step demand in flatness

We wish to have the closed-loop poles at $\lambda_1 = -1$, $\lambda_2 = -10$. This requires the characteristic equation

$$\lambda^2 + 11\lambda + 10 = 0.$$

Under closed-loop control, Eq. (8.1) becomes $\dot{x} = (A + BK)x$ where K is a feedback matrix to be determined.

   The characteristic equation is

$$\det \begin{bmatrix} \lambda - (k_{11} + k_{21}) & -(1 + k_{12} + k_{22}) \\ 3 - k_{21} & \lambda + 4 - k_{22} \end{bmatrix} = 0,$$

where k_{ij} are the elements in the K matrix.

If we now choose any matrix $(A + BK)$ that has the desired eigenvalues, this will allow K to be determined. For instance set

$$(A + BK) = \begin{bmatrix} -2 & -4 \\ -2 & -9 \end{bmatrix}$$

to yield

$$K = \begin{bmatrix} -3 & 0 \\ 1 & -5 \end{bmatrix}.$$

A check will show that the closed-loop eigenvalues have the values required.

8.6 Richalet's method for the control of complex multivariable industrial processes

This approach (Richalet *et al.*, 1977, 1978) can be summarised as follows.

(*a*) The process to be controlled is modelled off-line by an experimental method. The model consists of a set of stored impulse responses and has no equations, parameters or order.

(*b*) The model from (*a*) is built into the control algorithm i.e. it is stored and used on-line. There is no provision for automatic updating of the model or indeed any adaptive feature associated with the model.

(*c*) The aim of control is to force the process from its present known condition to the desired given condition along a smooth trajectory joining the two conditions.

(*d*) Process inputs are computed by use of the model in such a way that the process will be driven, so far as can be calculated, along the trajectory defined in (*c*).

The process model is built upon an assumption of linearity as follows.

The process has m outputs $y_1, \dots, y_m$ and r inputs $u_1, \dots, u_r$. Let time be discretised with the current time being the kth sampling instant. Then, approximately, the influence of the pth input u_p on the jth output y_j can be represented by

$$\sum_{i=1}^{N} a(i)_{p,j} u_p(k - i),$$

where the model has a memory of N time steps and the a are weighting coefficients.

Considering the effect of all the inputs on the output y_j yields

$$y_j(k) = \sum_{p=1}^{r} \sum_{i=1}^{N} a(i)_{p,j} u_p(k - i).$$

The sampling interval is chosen in conjunction with N such that $NT > \tau_{max}$ where τ_{max} is the longest settling time of any of the process responses.

It can be seen that the model is linear in the parameters — a considerable advantage from the point of parameter estimation.

Given a set of input—output data, the a coefficients are determined by a method known as structural distance minimisation. (This procedure is not fundamental to the overall control strategy since any other numerically efficient parameter estimation technique could be applied.) This procedure is in fact used by Richalet both for identification and control synthesis and it will therefore be described briefly.

8.6.1 Structural distance minimisation algorithm

(Here the simplest case — a single-input single-output model with quadratic criterion in a noise-free situation is used for illustration of the principles. See Richalet (1978) for the more general case.)

> Define the system output at time kT to be $y(k)$.
> The model output at time kT to be $\hat{y}(k)$.
> The system input at time kT to be $v(k)$.
> The n vector $[v(k-n), \ldots, v(k)]^T$ to be $u(k)$.
> The vector of process parameters, defined earlier, a.
> The vector of model parameters at time kT, $\hat{a}(k)$.
> The error between model and process at time kT, $e(k)$.

$$e(k) = \hat{y}(k) - y(k) = \langle \hat{a}(k), u(k) \rangle - \langle a, u(k) \rangle = \langle \hat{a}(k) - a, u(k) \rangle.$$

($\langle , \rangle$ denotes inner product.) Define a distance d in parameter space by

$$d(k) = \langle \hat{a}(k) - a, \hat{a}(k) - a \rangle.$$

Clearly d is positive definite, from the properties of inner products (or of quadratic forms).

If in a sequence $\{d(\cdot)\}$, $d(k+1) - d(k) < 0$ for all k then by the fixed point theorem (or if preferred by Lyapunov's second method) $d(k)_{k \to 0} \to 0$ and the sequence will converge to zero.

Define

$$b(k) = \hat{a}(k) - a,$$

$$b(k+1) = \hat{a}(k+1) - a = b(k) + \mu u(k),$$

where μ is a scalar coefficient governing convergence

$$d(k+1) - d(k) = 2\langle b(k), \mu u(k) \rangle + \langle \mu u(k), \mu u(k) \rangle$$

$$= 2\mu e(k) + \mu^2 \langle u(k), u(k) \rangle.$$

And the sequence will converge if we set

$$\mu = \frac{\lambda e(k)}{\langle u(k), u(k) \rangle} \quad \text{provided that } 0 < \lambda < 2.$$

For then $d(k+1) - d(k) = \lambda(\lambda - 2)\dfrac{e(k)^2}{\langle u(k), u(k) \rangle}.$

And the algorithm then becomes

$$\hat{a}(k+1) = \hat{a}(k) - \frac{\lambda e(k)u(k)}{\langle u(k), u(k) \rangle}$$

8.6.2 Determination of the model coefficients from plant data
Obtaining data rich in dynamic information requires very careful planning. The instrumentation of the plant, the injection of test signals both random and deterministic, and the duration and amplitude of such injected signals all need to be chosen. Unfortunately, the only criteria for decision involve a knowledge of the dynamics of the unknown process. Consequently, some preliminary trials and some simulation exercises will often be needed during the planning of the data collection trial. A number of different methods exist for the injection of test signals, but the safest method is to perturb the desired value settings of existing controllers since then the effects are quite predictable.

8.6.3 The control strategy
Any desired trajectory can be specified by the user but for many purposes a single first-order trajectory that commences at the current value of the process output and approaches the desired value asymptotically will be found adequate. Let the current value of the process output be $y(k)$ and the desired value be c. Then the desired value trajectory $y_d(\cdot)$ is given by

$$y_d(k+i) = \alpha y(k+i-1) + (1-\alpha)c, \quad i = 1, \ldots, n,$$

$$y_d(k) = y(k).$$

The coefficient α controls the rate of response.

8.6.4 Determination of the control inputs
The control inputs are determined, using the model iteratively, such that the model output follows as closely as possible the given reference input. These computed inputs are then applied to the actual process. The trajectories followed by the actual process will differ from the desired trajectories due to model errors and process disturbances, but the method seems to work well in practice despite this. Problems due to long-term model drift can be overcome by calculating control increments from incremental changes in the model outputs.

8.6.5 Advantages claimed for the method
On-line adaptation is not required. This removes the necessity for injecting test signals and in general makes the overall scheme more acceptable in a typical industrial plant.

The reference (Richalet, 1978) gives brief case histories of three successful implementations (PVC plant, oil refinery, electricity generation steam plant). What is impressive is that each implementation is running routinely on a full-scale plant and making well-documented savings.

Chapter 9

Computer control methods

9.0 Introduction

Although many earlier generation process computer systems will persist for a number of years, it is safe to assume that new systems will be increasingly microprocessor based. Accordingly, this chapter concentrates on systems for achieving process control through microprocessors, with emphasis on recently available commercial systems.

9.1 Multilevel (hierarchical) control configurations

The archetypal multilevel configuration is the conventional pyramidal management structure. Since a computer control system is designed to assist management to achieve its objectives, it is not surprising that the control configuration shares many of the characteristics of the management system.

Consider the management hierarchy required for a completely manual complex process. At the lowest level, the process operator receives input information from measuring sensors and direct observation. His goals are passed down from an upper level of the hierarchy. His outputs are corrective actions to the process to achieve the given goals and information on achievements which is fed back to an upper management level. The actions taken are to a large extent feedback actions, the time constants involved are relatively short and the goals are short term.

By contrast, upper levels of the hierarchy are essentially concerned to achieve an overall long-term plan. Taking account of information arising from the lower levels in the hiearchy, together with external information, decisions are made which affect the goals and even the configuration of the lower levels. At the upper levels, the time constants are long and most decisions are based on predictive techniques.

The distributed control systems described in the next section will be found to be ideally structured for the implementation of hierarchically inspired control systems.

9.2 An introduction to distributed process control systems

Distributed process control systems comprise *local control stations*, capable of autonomous operation, linked by a *data highway* and co-ordinated by a *central facility* (Fig. 9.1).

9.2.1 The advantages of the distributed approach

Early process control schemes (generation 1) used local analogue controllers. Since they operated independently, in parallel, the resulting system was inherently reliable. However, it was difficult to obtain a satisfactory overview of a large system and it was impossible to apply an integrated control strategy to the complete process.

Next on the scene came large process control computers (generation 2). Their centralised operation made sophisticated integrated control available. Disadvantages were: their complex software structures required extremely time-consuming efforts at all stages (design, programming, commissioning, maintenance, extension); some form of back-up was usually necessary (spare computer on stand-by or provision of a stand-by, generation 1, analogue system).

The nominal advantages of distributed control are:

(*a*) The autonomy of the local controllers allows the reliability and simplicity of generation 1 schemes to be attained without special provision being made.
(*b*) Intercommunication via the data highway allows integrated control of the process as in generation 2 systems.
(*c*) The complete system can be configured, commissioned and operated without the need for extensive software effort. Algorithms called from manufacturer-supplied PROM's are interconnected by programs resident in the local random access memory. This interconnection by program ('softwiring') is often achieved using a control-oriented high-level language. The system is readily expanded and modified.

The local controller is microprocessor based. A typical unit is shown in Fig. 9.2. The system executive is stored in PROM, as is a library of (typically) fifty functions and algorithms (such as signal conditioning and PID algorithms) that the process engineer can use as building blocks. The building blocks are interconnected by a RAM resident program using a control oriented high-level language resembling BASIC language. The security of operation of the control loops depends on interconnection through programs stored in volatile memory. In other words, in the event of a power failure, the loops are broken and the programs are destroyed. To overcome the problem, at least one on-board long-life battery provides power back-up for the RAM memory.

Input/output modules can be plugged in according to the needs of the application. Communication with the data highway is provided through a data link module.

A local operator's display and input panel, and provision for process engineers to test and modify parameters complete the basic specification.

Local controllers often carry out self-diagnostic checks with automatic change-over of equipment in the event of a fault. The degree of autonomy of the local

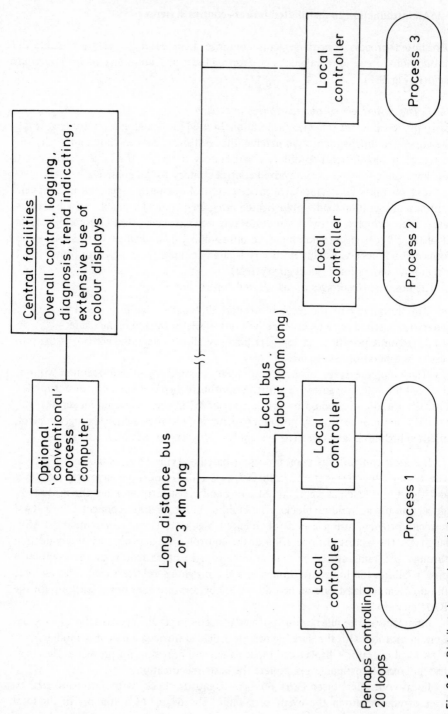

Fig. 9.1   *Distributed control — typical system*

controller is a key factor to consider since the more autonomous the controller, the less the dependence on a possibly vulnerable data highway. However, high local autonomy may make it difficult to achieve central overall control of a large process. As a general rule, processes that are only loosely coupled together should be controlled by highly autonomous local controllers.

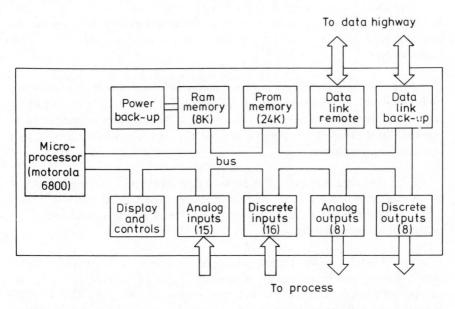

Fig. 9.2   *A typical small local controller (the Micon unit of DRD Mess- und Regeltechnik GmbH)*

The data highway is typically a serial link capable of transmitting data at up to one million bits per second. With many active devices on the data highway, the protocol for assigning intercommunication between two particular devices becomes fundamental to the operation of the whole system. Such control may be completely centralised or control of the bus may pass from one device to another (the so-called *flying master* system). Some systems use a local bus to link devices in groups when they are tens of metres apart. The groups are then interconnected by a second long-distance bus. The complete data highway may be duplicated as a measure to increase reliability.

The central facility will vary considerably with user requirement but, in any event, it will be built up from standard modules and parametrised by similar methods to those used for the local controller. In a large system, there will be extensive display outputs. These will offer such facilities as historical trend data and condensed overall performance data, in ergonomically attractive formats using colour display techniques.

Overall control will be achieved by vendor supplied algorithms, as in the local controller, supplemented if necessary by custom programs unique to the application.

Provision is often made for a conventional process computer to be interfaced with the central facility.

9.2.2 Some suggestions for the main characterising variables that allow broad comparisons to be made between rival systems

In order to allow some appreciation of the relative merits of alternative systems, the following suggestions are made for parameters that would largely characterise a particular system.

9.2.2.1 The local controller: Degree of autonomy. Maximum number of control loops. Minimum sampling rate. Range of provided algorithms. Ability to perform sequencing operations. Method of programming. I/O modules, types and maximum number. Level of operation possible without the data highways. Degree of local man—machine interaction.

9.2.2.2 The data highway: Physical type (e.g. fibre-optics). Type of configurations (e.g. star). Protocol for control of information transmission (e.g. central master or flying master). Note that the data highway can be duplicated in many systems to increase reliability.

9.2.2.3 Overall aspects: Largest manageable system size. Provision for central alarms, logging and man—machine interaction. Types of displays. Reliability enhancement through self-diagnosis and re-allocation of functions. Ability of the systems to be built up, maintained and expanded by engineers rather than software specialists. Ability of the system to mirror a natural hierarchical command structure that can be appreciated by plant personnel. Compatability of the system with existing equipment to ease the transition from generation 2 to generation 3 operation.

9.2.3 A brief description of some representative systems

(Further details of these systems should be sought in the manufacturers' literature.)

The Honeywell TDC 2000 system (Totally Distributed Control) was announced in 1975 — as such it was one of the first systems to become commercially available. The Honeywell system will be described in some detail and then further available systems will be described more briefly.

9.2.3.1 The Honeywell TDC system

The Basic Controller (BC): Contains the General Instruments CP1600 16-bit microprocessor and is designed for continuous control of conventional analogue feedback loops. The controller takes in up to 16 analogue signals, processes the signals through 28 ready-made algorithms stored in PROM, and produces 8 analogue outputs. Each output variable is scanned and each output variable is updated at a fixed rate of 3 times per second. All plant input/output signals are in the form of

4–20 mA currents. The algorithms are selected and parameters set from a basic operator station on the data highway. The basic controller is powered by 24 volt supply with battery back-up.

The Reserve Controller Director (RCD): This is the central unit of the Uninterrupted Automatic Control System Concept (UAC). The RCD detects the failure of any of eight controller files, informs the operator of the failure and switches in a reserve file with correct parameters — all within one second of the failure occurring.

Process Interface Units (PIU's): They allow:

(*a*) Efficient data acquisition from plant sensors into the data highway.
(*b*) Switch type signals originating from a supervisory controller on the data highway, to be output to the plant. (Recall that the Basic Controller cannot undertake discontinuous duties so this facility is needed for logic operations during start up, etc.).

Data Highway (DH): Three branches, each up to 5000 feet long and consisting of standard co-axial cable, connect into a single HTD to interconnect up to 63 devices (up to 28 to a branch). Data are transmitted serially as 31 bit words at 250 000 bits per second. Each device is 'hung-on' to the highway via a standard interface. Each branch of the highway can be duplicated, to give a *Redundant Data Highway* (RDH), with the HTD performing switchover between primary and back-up cables. The primary and back-up cables will usually be separated physically (one overhead and one underground perhaps) to improve reliability. A number of accuracy checks are built into the communication protocol including the requirement for a device to echo each signal that it receives to allow its accuracy to be checked at source.

Highway Traffic Director (HTD): The HTD controls the protocol of intercommunication along the data highway. It contains duplicate logic to cover both primary and back-up operation with switchover to a back-up data highway cable.

A Basic Operator Station (BOS): This is a 19 inch colour VDU with keyboard connected into the data highway. In a typical arrangement three identical BOS are sited adjacently so that all three can be seen by the same operator. One is usually showing a plant overview, one is switched to group displays and one is dedicated to alarm duties. A hard-copy unit will normally be arranged to record all alarm conditions that occur. Flexible, comprehensive monitoring (up to 200 points per second) is a feature of the system. The operator interacts with the system (calling up different types of displays or changing parameters or configuration in the basic control loops) by means of push-buttons.

The Supervisory Operating Centre (SOC): This connects to the data highway and contains a Honeywell 4500 process computer with a further fifty ready-made algorithms. The SOC provides more complex, more comprehensive facilities and a more

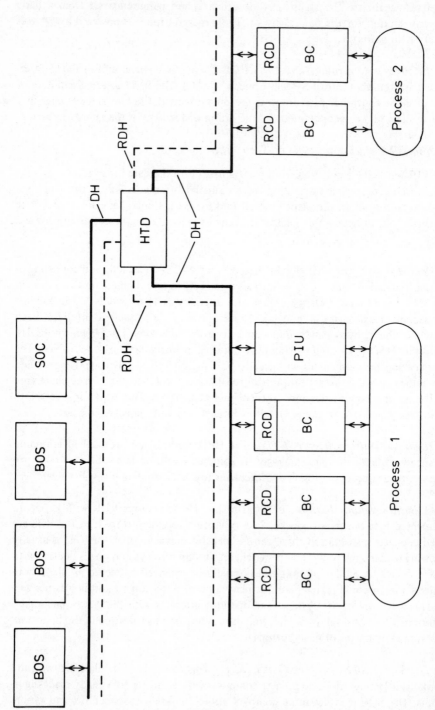

Fig. 9.3   *Honeywell system TDC 2000 – typical configuration*

long-term overview of the plant. Displays available at the SOC include 'Dynamic Graphics' so that (for instance) a tank can actually be seen to be emptying on the screen as a true mirror of the plant.

Figure 9.3 illustrates the layour of a medium-size system. Much larger systems are possible, culminating in what Honeywell calls 'Total Control'. Conversely, quite small systems (say 5 or 6 loops) can be made up at low cost using one Basic Controller and without a Data Highway. In such a case, the Basic Controller is set up and the operator interacts through a Data Entry Panel (DEP), not shown in the figure, since its function is covered by the Basic Operator Station in the system.

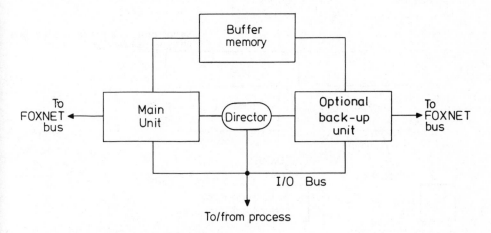

Fig. 9.4 *Foxboro's local controller MICROSPEC*

9.2.3.2 The Foxboro System (Figs. 9.4 and 9.5)
Foxboro have developed their distributed system, based on MICROSPEC controllers, and a bus system, FOXNET, within the context of the overall SPECTRUM system. This system contains the process computer systems Fox 1 A and Fox 3, VIDEOSPEC (a VDU based monitoring system) and SPEC 200 (an analogue-based control system connectable into digital systems).

Up to ten local controllers may be connected to a LINKPORT through which they may communicate in multi-master mode. Up to ten LINKPORTS may be interconnected by a serial data link. Existing systems such as Fox 3 may be connected into a LINKPORT — an important compatability allowing extension of earlier systems making use of latest technology.

9.2.3.3 Siemens System – Teleperm M (Fig. 9.6)
Three micro-based local controllers are provided by Siemens allowing for small (15 loop), medium (120 loop), and large (200 loop) operation with the largest controller (type AS 230) having considerable storage of up to 64 K words PROM and 320 K words RAM.

The local controllers are interconnected by a local bus over distances up to

100 metres. Further interconnection is made possible through a bus-coupler to the long-distance bus, with a range of 4 km.

9.2.3.4 Fischer and Porter System DCI-4000 (Fig. 9.7)

The DCI-4000 (Distributed Control Instrumentation) system uses up to four distributed control units (DCU's), connectable to a central operators panel. Figure 9.7 gives some salient features of the system.

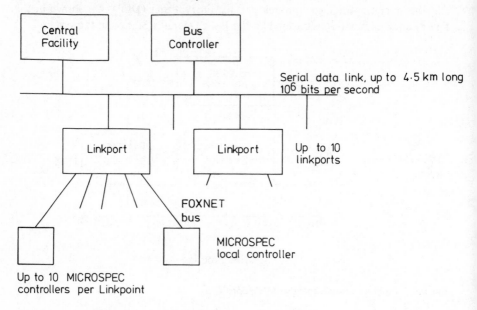

Fig. 9.5   *Foxboro system – overall configuration*

The local controller (DCU) is based on an LSI 11. One DCU is designated as a back-up (see digram). If self-diagnosis shows that one of the working DCU's is faulty, the back-up unit is automatically substituted to maintain operation. The back-up DCU therefore needs to have the capacity and access to data to allow it to undertake any of the four possible functions that it might be called on to perform.

A total of 16 DCU's can be connected into the system as indicated in Fig. 9.7, to form a classical hierarchical structure.

9.2.3.5 Kent System P4000

The Kent system utilises three basic types of local controllers:

(a) A simple analogue PI controller ('Basic Controller').
(b) A more sophisticated 'Intermediate Controller' that represents an extension of (a).
(c) A micro-based 'Advanced Controller'. The structure of this is outlined in Fig. 9.8.

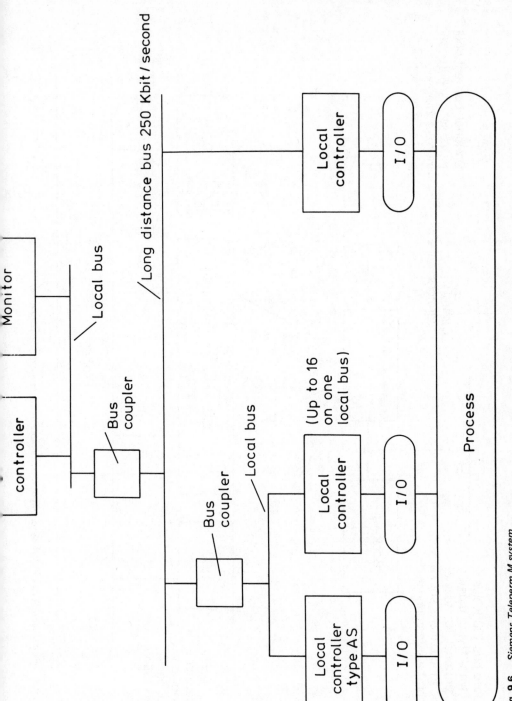

Fig. 9.6   *Siemens Teleperm M system*

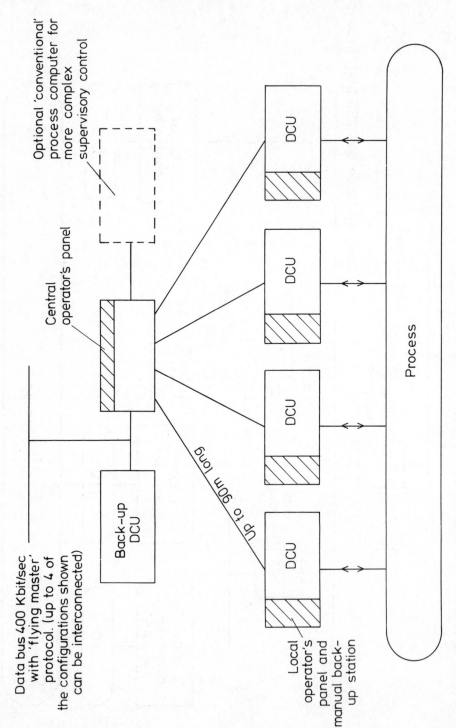

Fig. 9.7   *Fischer and Porter DCI 4000 configuration capable of handling 256 control loops*

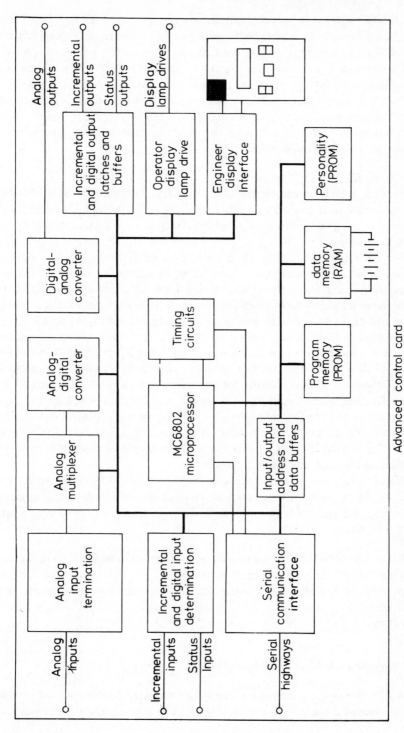

Fig. 9.8   *Kent system P4000, advanced controller card*

The philosophy is to use one controller for one loop to obtain high inherent reliability. Up to 500 controllers can be connected to a central monitoring and control station that can include one of the Kent process control computers K70 or K90.

Two physically different arrangements are available: 'Discrete Instrumentation' and 'Split Architecture', although the functional arrangements are the same for both. It is a matter of customer taste and industry tradition which physical arrangement is specified.

A number of different groupings and data transmission configurations can be achieved for different applications.

9.2.3.6 The SCIDAC System

SCIDAC is a data acquisition and control system developed by SCICON (Automation Division). It consists of a set of hardware and software building blocks that are used to construct a process control or industrial control system to a customer's specific requirements.

Systems start at the level of a single microprocessor-based unit and range through small distributed networks controlling the operations of a single manufacturing site to large telemetry systems as used in the oil, power and water industries. Each computer unit is based on INTEL hardware using multiple 8086 processors. A single unit can handle up to 50 analogue loops and some 400 digital signals.

A typical SCIDAC system for a process control application utilises several front-end units connected to a central supervisory unit using separate point-to-point serial links. All data acquisition, monitoring and control, both continuous and sequence, is performed by the local units. The central unit provides all the standard supervisory facilities such as colour mimic display, trend recording and plotting, alarm annunciation and logging, and report generation. Operator control over the local units can be exercised at work-stations attached to either the local or central equipment. The central unit also holds the process data-base, production schedules, machine settings, job specifications and so on.

A wide-ranging set of facilities is provided for the user to reconfigure his equipment on-line to cope with the inevitable changes in plant, instrumentation, control techniques and operational practices that will occur during the life of any industrial control system.

The systems that have been described above represent a selection from the 'large' systems available. It has to be emphasised that these systems can, because of their modularity, be applied to quite small problems. Conversely, there are many, nominally small, systems on offer that, by interconnection, can control large complex processes.

9.3 Programmable logic controllers (PLC's)

Programmable logic controllers are microprocessor-based devices designed to replace sets of relays for control of sequencing in industrial processes. The operation of a programmable logic controller is governed by the program whereas, for a set of relays, time-consuming and difficult to modify hard-wiring is required. Of course, logical functions can be effected by almost any computer. The advantage of programmable controllers is that, being designed for a single purpose, they are rapidly implemented by personnel who, though familiar with process requirements, are not competent in real-time computer software.

Buschart and Hohlfeld (1978) give figures based on experience in the chemical industry, showing that for small systems, programmable logic controllers may be up to 40% cheaper than conventional relay systems — these figures include installation and programming costs. The smaller commercial programmable logic controllers store their operating program in PROM. The facilities are basically AND/OR gates, timers and counters. Programming is often performed through a separate portable computer which, being relatively expensive, may be borrowed from the vendor. For applications where the program will have to be modified from time to time, the ease with which this can be accomplished will need to be examined carefully. PLC's are offered by over thirty manufacturers, see Hickey (1978). Memory sizes range from 1 K upwards, often being expandable incrementally. PLC's can have up to 2000 inputs. Programming by relay ladder diagrams is offered by most manufacturers although Boolean languages and other options may be offered additionally.

Chapter 10

A selection of applications

10.0 Introduction

A few years ago, control applications were largely restricted to the traditional areas: Aerospace, Chemicals, Petroleum, Power, Metals and the like. The range of applications has recently increased enormously to embrace such diverse areas as: control systems for car engines, systems for automatic unloading of oil tankers, energy management systems for large buildings, sophisticated mechanical handling systems and automatic exposure and focussing for cameras.

The newer application areas have not, from the point of view of control theory, posed particularly novel or illuminating problems. Accordingly, the applications to be described in this chapter are largely from the traditional areas.

The list of references should be consulted for further reading. Really good applications papers are unfortunately rare, although a consistently useful source is the German journal *Regelungstechnische Praxis*. (Its sister journal *Technische Messen* deals with measurement problems.)

Some of the applications are based on the author's experience in the steel industry. For further background on steel processes from a control point of view, see Leigh and Williams (1972).

10.1 Control of temperature

Temperature control, admittedly often of a primitive kind, is to be found in virtually every building. Many industrial processes operate at high temperatures and both quality control and fuel economy considerations demand tight control of temperature. Despite this, temperature control receives little emphasis in most control texts. Most aspects of temperature control are, of course, common to all control problems. Here I have outlined only those problems arising from the essential asymmetry of most temperature control systems. The basic difficulty is that in a temperature control system, temperature rise is produced by (controllable) energy input while temperature fall is produced by heat losses which are not only uncontrollable, but which vary significantly and to an extent unpredictably.

Consider the very simplest closed-loop temperature control system where a temperature x is to be maintained as near as possible to a constant desired temperature x_d. What heat input should the process receive when $x = x_d$? Here we have zero temperature error but the heat input to the process must be non-zero except in the special case that x_d is the ambient temperature.

Clearly, the answer must depend on a number of factors including x_d, the degree of insulation of the process and the ambient temperature.

Commercial temperature controllers usually have an output as shown in Fig. 10.1.

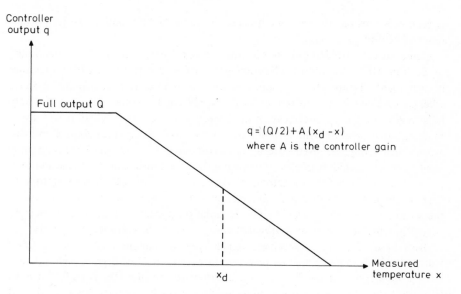

Fig. 10.1   *The characteristic of a temperature controller*

Thus, arbitrarily, the controller puts out half its full available output where $x = x_d$. What is the effect of this? In general the answer is complex but below we assume that the controller feeds a first order system described by the equation

$$\frac{dx}{dt} = \frac{1}{c}[k_1 q - k_2(x - x_{\text{ambient}})],$$

where k_1 and k_2 are process constants, and c is a thermal capacitance term.

For equilibrium $dx/dt = 0$.

Putting $dx/dt = 0$, $x_{\text{ambient}} = 0$ and solving the equations above simultaneously yields the temperature error in the steady-state as

$$x_d - x = \frac{k_2 x_d - k_1(Q/2)}{k_1 A + k_2}.$$

This equation shows at once that there will be a steady-state error except in the special case that the numerator is zero. Such steady-state errors tend to be large and if integral action is incorporated for their removal, this tends to require a large degree of integral action with degradation of the transient performance of the controlled system. For systems where x_d is restricted to lie in a narrow band, the value of Q should be chosen carefully. For systems where x_d moves over a wide range, a four-term temperature controller has been suggested (Leigh, 1977*a*).

10.2 Control of fuel-fired furnaces and enclosures

A section is devoted to this topic because of the importance and widespread occurrence of heating processes of this type.

Some special problems of temperature control have already been mentioned in Section 10.1. An additional problem occurs in connection with temperature measurement. Temperature sensors measure the temperature at a point in space whereas the temperature of interest is almost always the temperature in a relatively large region of space. The location of the temperature sensor where it will give a representative measurement is fundamental to good temperature control and this aspect must always be given special consideration. If thermocouples are used, they need to be protected both against damaging gases and mechanical damage and such protection can introduce measurement time constants of sufficient magnitude to degrade the control system performance. Incorrect temperatures will be obtained if flames impinge on the thermocouple. If radiation pyrometers are used, the emissivity of the material to be measured will always need to be considered carefully.

Because of the highly nonlinear dependence of radiation on temperature it works out that running a high-temperature furnace even a few degrees higher than is strictly necessary can result in a large addition to the fuel bill. Therefore it is an important part of the control scheme to decide on the correct desired value of temperature for each schedule.

A second, related problem, is to ensure that the heating process is carefully matched to the processes that it feeds so that, on the one hand it does not delay these processes by failure to satisfy their needs nor does it waste energy by overproviding. This matching needs special care and has a considerable cost benefit where the processes are subject to intermittent stop—start conditions and/or variable throughput.

Thermal efficiency is influenced greatly by correct combustion. This involves:

(*a*) ensuring the correct ratio of air to fuel in the burners;
(*b*) ensuring that, so-called secondary air is not drawn in to the furnace or enclosure. This must be achieved by adequate furnace pressure control.

Correct combustion can be monitored most efficiently by gas analysis of the waste gases. Sufficient air must be admitted to burn all the fuel but excess air must be avoided. We can imagine that excess air takes no part in the combustion process, but that it is heated up to the working temperature of the furnace and then leaves, taking with it sensible heat. Provided that the analysis of the fuel is known reasonably accurately, the state of combustion and the avoidance of excess air can be monitored very simply and cheaply by continuously measuring carbon dioxide in the waste gases. For instance, if the fuel were pure carbon and air is supposed to consist of 20% oxygen and 80% nitrogen, then with complete combustion and no excess air

$$C + O_2 + 4N_2 \rightarrow CO_2 + 4N_2 .$$

Thus there should be approximately 20% by volume of carbon dioxide in the waste gases with full combustion.

Either an excess or deficiency of oxygen will reduce the percentage of carbon dioxide in the waste gases. In the first case by simple dilution and in the second case because carbon monoxide begins to be produced instead of carbon dioxide.

Fuels do not consist of pure carbon and the calculation needs to take this into account. As a rough guide, furnaces burning fuel-oil or gas should have a carbon dioxide content of about 14% by volume in the waste gas. Oxygen can be mesaured continuously on a routine basis by a zirconia sensor and this gives a more direct indication of thermal efficiency in return for increased investment.

It is not essential to use an automatic loop to compensate and control combustion based on gas analysis, since correct combustion can be maintained fairly easily by periodic manual adjustment. Typical 'standard controls' for a fuel-fired furnace are shown in Fig. 10.2.

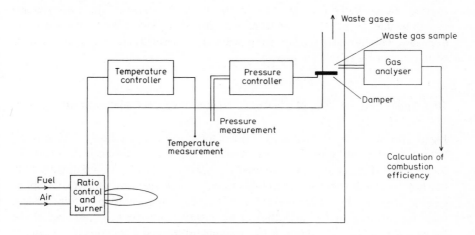

Fig. 10.2 *Typical control system for a simple fuel-fired furnace*

The burners have automatic maintenance of air/fuel ratio built into their design. The two control loops are quite conventional in their operation.

One of the most difficult practical aspects of furnace control is to maintain good thermal efficiency over the wide range of fuel inputs that may obtain in practice. For instance, when the furnace is at the required temperature the fuel input may be only five per cent of maximum, with the corresponding amount of air. The furnace pressure control system must operate successfully from this condition to full fuel input to prevent ingress of air and hence to maintain high thermal efficiency.

Furnace pressure control can prove difficult in a number of practical ways.

(*a*) Gusts of wind acting on chimneys cause rapid pressure transients that no ordinary system of furnace dampers can hope to compensate.

(*b*) Furnace pressure is necessarily measured differentially with an outlet somewhere being regarded as datum pressure. Since very low pressures are involved it is difficult to site this datum outlet, since few locations have a really constant air pressure.

(*c*) Buoyancy effects are remarkably large in high-temperature furnaces and pressure control is only really possible along one chosen horizontal plane. Some furnaces are constructed deliberately on a considerable slope and the attainment of good pressure control in these is all but impossible.

(*d*) One problem which I encountered in practice in a furnace pressure control system and which I have never analysed scientifically is as follows.

Suppose that, with the system operating satisfactorily, the desired value of furnace pressure is increased only slightly. It happens that, with the increased pressure, some furnace gases begin to escape from the enclosure by 'unofficial' routes. (These gaps in the furnace enclosure are always there — if they were not there would be no need for pressure control.) This causes a loss of furnace pressure and the control system operates to close the exit damper. The effects are cumulative and culminate in the dampers being completely closed with all the gases of combustion being discharged spectacularly through gaps in the furnace enclosure. The converse also applies in that, if the desired pressure is decreased somewhat, air is drawn into the enclosure. This increases the internal pressure and eventually the exit dampers move to and stay in the fully open position.

10.3 Temperature control of a jacketted reactor using a cascaded controller

Figure 10.3 shows a jacketted reactor where the temperature of the agitated contents is to be controlled by the hot-water flow into the jacket. Although the measured temperature of the contents can be used to operate in a feedback loop directly onto the hot-water valve, better control can usually be obtained by using two loops in cascade as shown. Jacket temperature is controlled in an inner loop with the desired value for this loop being the output of the master controller. The response of the jacket must be fast compared with the response of the vessel contents if the cascade approach is to be worthwhile. The inner loop is tuned first and the outer loop is tuned with the inner loop operative.

10.4 Control of exothermic processes

When a passive material is heated, the heat losses (being dependent on conduction, convection and radiation) are difficult to calculate accurately, but nevertheless the curve of heat losses against temperature is known to be of the form given in Fig. 10.4. Radiation, with its highly nonlinear characteristic, eventually dominates as temperature rises. For a particular heat input such processes are highly self-regulating. Chemical reactions are classified into:

(*a*) Endothermic reactions – those where the reactions produce a cooling effect. Heat needs to be put into these processes to force the reaction in the right direction, and no specially difficult control problems arise.

(*b*) Exothermic reactions – those where heat is produced by the reaction. Heat needs to be removed from these processes as they proceed. These processes pose difficult control problems.

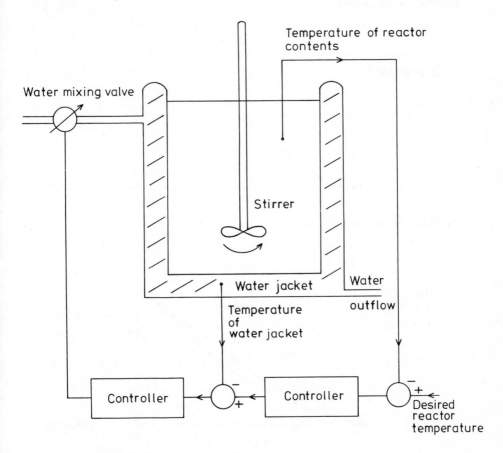

Fig. 10.3   *Cascade control of a jacketted reactor*

Precise temperature control is necessary to regulate the rate of reaction, to ensure plant safety and to satisfy quality assurance requirements.

Exothermic processes are in general not self-starting. Initially, sensible heat needs

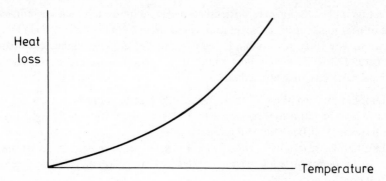

Fig. 10.4   *Heat-loss curve*

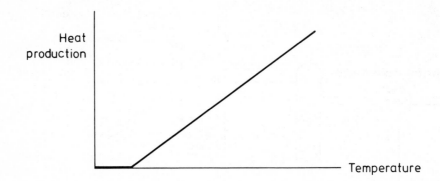

Fig. 10.5   *Heat-production curve*

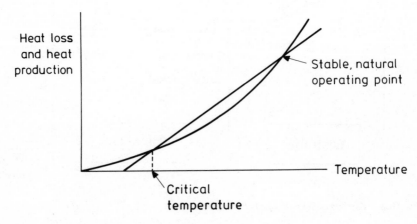

Fig. 10.6   *Figures 10.4 and 10.5 combined*

to be supplied to the process. Once a critical temperature is exceeded, the heat of reaction exceeds the heat losses and the temperature rises. The rate at which a chemical reaction takes place is highly temperature dependent and the process, now with inherent positive feedback, may run away unless the cooling system is adequate. A control system for a batch-working exothermic process is thus required first to initiate the process by adding heat, and then to control the temperature by extracting heat. A curve showing the heat production from the process might be as shown in Fig. 10.5. This, in combination with the natural heat-loss curve of Fig. 10.4, yields the overall behaviour of Fig. 10.6. Here the critical temperature is seen as that temperature where the heat produced begins to exceed the natural heat losses and forced cooling becomes necessary. Notice that the critical point is not an inherently stable operating point for the process. To see this, assume a small positive temperature perturbation. This causes an overall heat gain, and the temperature starts to run away. The point marked, natural operating point, is a stable point since temperature deviations tend to be self-correcting there. However, this temperature will usually be much too high for safe operation of the process. Control is achieved by manipulating forced cooling. In effect, a family of cooling curves is available, each one being able to fix a new stable operating point.

10.5 Control of pH

pH is a measure of hydrogen ion concentration or, roughly speaking, of the degree of acidity of a solution. pH needs to be controlled in many industries in addition to the obvious chemical industries. Legal requirements mean that discharges into rivers must not be too acidic and the waste from various acid baths needs to be neutralised with minimum alkali give-away, before it is discharged. The pH of water gathered in upland reservoirs is very acidic, with a pH of about 4. Before use as drinking water it is brought to a pH of about 8 by the addition of lime to the service reservoirs. If the water is stored in the open in a city, or what is equivalent, recirculated in industrial cooling system, the pH soon falls due to the presence of atmospheric acids, and corrosion problems can result unless steps are taken to keep the pH at a high enough value. Water is an unbuffered solution from the point of view of pH control. This means that the pH is extremely sensitive to added acid or base. Strong solutions on the other hand are heavily buffered, and change their pH only little on the addition of acid or base.

In addition to the variations in behaviour due to buffering, there is an important fundamental nonlinearity in the process. Consider any particular solution at a very low pH and examine the change in pH as a strong base is added. The resulting curve (the titration curve) is highly nonlinear, as shown in Fig. 10.7. If there is uncertainty about the degree of buffering and if, due to poor mixing, there is significant deadtime in the control loop, then these factors, together with the highly nonlinear characteristic of the titration curve, make pH control quite difficult. In many applications, the pH is to be controlled at or near to a value of 7. The process gain is here at its highest. For stability, the controller gain will need to be set to a low value, but this will give a poor performance whenever the pH is far removed from its desired value. The solution is to use in the controller an approximation to the inverse of the titration curve so that, acting in series, the two nonlinear functions combine to give approximate linearity. This, together with well-designed active mixing of the solutions, should normally allow pH control to be achieved without special difficulty.

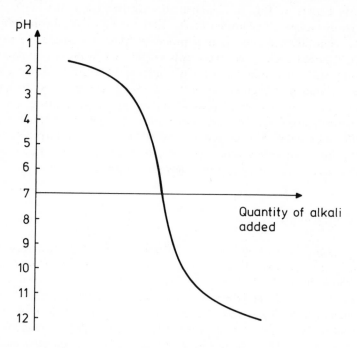

Fig. 10.7   *A titration curve*

10.6 Blending line control

Blending lines are designed to mix together two or more liquid components, in exact proportions, on a continuous basis. Blending lines find wide application in the petroleum, chemical, paper and food industries, where they are displacing traditional methods of mixing in batches.

Consider the simplest application where two streams are to be blended in specified ratio. Each pipe-line is fitted with a turbine flowmeter that produces a train of pulses at a rate dependent on flow rate. The desired flow rate for each stream is expressed as a pulse rate and the comparator between desired and actual flow rate is an up/down digital counter whose state is therefore the integral of the error between desired and actual flow rates. The desired value pulse rate originates from a microprocessor that maintains the required ratio between components as well as controlling the overall flow rate. Although the system is essentially continuous, the program will usually allow a batch of a specified volume to be dispensed. Some blending systems incorporate a pacing facility where, should the flow of one of the component streams fall below the desired value, then the other stream will be held back also in the correct ratio. Notice that, if a control loop is designed to control ratio directly, then this will result in nonlinearity because of the division involved.

10.7 Viscosity control

Davis and Smith (1977) describe an interesting application for the control of fuel oil viscosity in a 600 gallons/minute process supplying a power station. After rejecting,

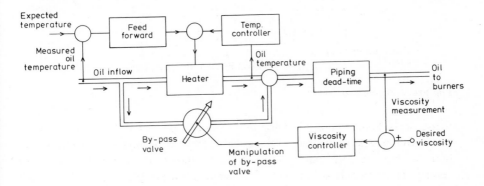

Fig. 10.8   *Viscosity control system*

on grounds of poor performance, a simple closed loop in which viscosity is the measured variable and the oil heater is manipulated, they presented several alternative schemes and evaluated them by simulation. Their recommendations contain some generally useful lessons. Figure 10.8 shows the final scheme. The temperature control loop is the main control loop. It has little dead-time and relies on well-tried hardware. The viscosity controller makes rapid adjustment to viscosity by manipulating the by-pass valve – note that this feature, together with its piping, has to be built to allow the viscosity controller to operate. Viscosity measurement is rather unreliable but if this loop fails, the oil viscosity will still remain within a usable region through the actions of the temperature control loop. The feedback loop is justified only if the oil flow rate is expected to fluctuate very significantly.

10.8 Control of the sinter plant (Fig. 10.9) [see Rose and Radmanesh (1981)]

The sinter strand is basically a continuously operating wide conveyor belt carrying a bed of iron ore mixed with coke. Flames impinge on the surface of the bed as it passes through the ignition hood. Combustion progresses downward, assisted by suction from beneath, as the conveyor advances. The control aim is for the process to operate at maximum throughput whilst satisfying quality requirements in the sinter produced. In particular, the lumps of sinter must have a high mechanical strength – if not they will disintegrate in the tower-shaped blast furnace and adversely affect

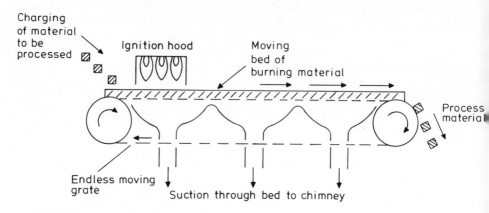

Fig. 10.9   *The sinter plant*

that process. Some lumps inevitably disintegrate to powder at the end of the sinter plant. This powdery material, perhaps 15% of the throughput, is not transmitted onto the blast furnace, but is returned to the start of the sinter plant for reprocessing. The manipulable controls of the sinter plant are bed depth, strand speed, coke/ore ratio and initial added moisture. To understand the control problem it is necessary to understand the concept of 'burn-through point'. Considering a narrow element of the strand, the material can be grouped into three categories: burned, burning and unburned (see Fig. 10.10). The first point along the strand where nothing remains except burned material is called the burn-through point. The control objective is to operate such that the burn-through point coincides with the end of the strand. The permeability to air of the bed can be measured by pressure drop techniques, and a control loop can be used to manipulate initial added water to maximise this permeability. The main problem with the air permeability control loop is that the curve relating air permeability with moisture content passes through a maximum and the control system needs to 'know' on which side of the maximum it is operating, otherwise the sign of the feedback will be incorrect. For the measurement of mechanical strength, magnetic sensors have been devised that measure magnetic permeability of the finished sinter. Tests have established that magnetic permeability is in a 1:1 relation with mechanical strength, although the particular relation has to be established for each type of raw material. Automatic control of mechanical strength is then achieved through manipulation of the coke/ore ratio. There is a dead-time of several minutes in this control loop, being the time for a change in the coke/ore ratio to traverse the strand and affect the measuring sensor. The returned fines complicate the response and also lead to logistics problems. If

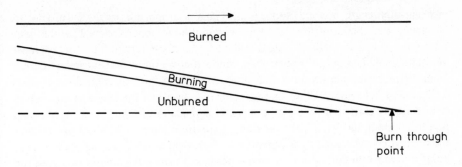

Burned

Burning

Unburned

Burn through
point

Fig. 10.10   *The sinter plant — a section through the bed*

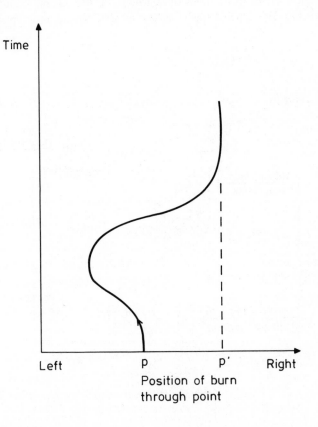

Fig. 10.11   *The non-minimum phase behaviour of the burn-through point in response to a step increase in bed thickness*

mechanical strength becomes too low, too large a proportion of fines will arise to be recirculated. Some of the fines can be stored in a hopper designed for that purpose but its capacity is limited and, in the long term, a returned fines balance must be established. The burn-through point cannot be measured, but it can be inferred from thermocouples in the air streams that have passed through the bed.

Transient behaviour of the plant is interesting. Consider the case where the plant is in a steady state and then a step increase is made in bed thickness. As the step progresses along the strand, the air flow distribution alters with those parts still at the original bed thickness receiving more air flow than usual. Thus, the burn-through point moves to the left while the step is progressing along the strand. However, the new steady state for the strand with increased bed thickness will clearly have its burn-through point p' to the right of p, where p is the original burn-through point (see Fig. 10.11). This type of behaviour is typical of non-minimum phase processes.

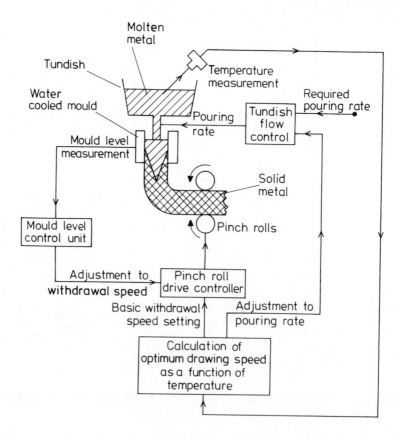

Fig. 10.12   *Control of continuous casting of steel*

10.9 Control of continuous casting

In continuous casting, a stream of molten metal is poured into a bottomless water-cooled mould and, with its exterior solidified, is pulled through by pinch rolls to produce a metal strand of relatively small cross-sectional area suitable for subsequent processing. A typical machine will simultaneously cast eight strands of rectangular cross-section for subsequent rolling into strip.

The chief control problems are concerned with:

(a) the flow rate of molten and solidified metal;
(b) the temperature of the material;
(c) optimal start-up and shut-down.

A number of difficult variables have to be measured including the levels of molten metal. Figure 10.12 shows one possible configuration for control of the process. The aim is to achieve maximum throughput rate at all times. This is calculated by an on-line model from data on molten metal temperature and coolant rates. The molten metal levels are controlled to balance the flow rate.

Optimal start-up and shut-down is concerned with control of process logistics and with producing in each batch exactly the right lengths of material needed for subsequent processing — any excess would have to be recycled as scrap.

The control aims are achieved in an integrated scheme.

10.10 Control of the flatness of steel strip

10.10.1 Definition of the problem
Many products, such as car bodies, refrigerators, cans and transformer laminations are made from cold-rolled steel strip. The material is rolled at high speed to exacting tolerances. We discuss here the important problem of non-flatness arising from non-uniform elongation during the rolling process (see Fig. 10.13). Such non-uniform elongation arises, either becuase the rolls are not perfectly parallel, or because the incoming material is not perfectly symmetrical.

In strip rolling, the rolls need to have a small diameter, otherwise they cannot 'bite' the material and the operation is inefficient.

The rolls can only be supported at their ends and the large forces produced during rolling cause roll bending in a vertical plane. This leads to a non-uniform strip velocity profile across the width and hence to non-flat strip.

To counteract this effect two mechanical design measures are adopted. The work rolls are cambered (i.e. of greater diameter in the centre) and are supported by large diameter support rolls (Fig. 10.14).

These measures give perfect compensation only at one particular roll separating force and any departures have to be compensated by automatic control loops.

10.10.2 *Economic evaluation*

Extensive analyses revealed that improvements in the degree of flatness of steel strip were economically highly desirable, particularly in the buyers' market that exists for this type of product.

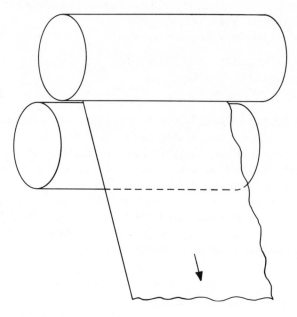

Fig. 10.13   *Non-uniform elongation during strip-rolling*

10.10.3 *Measurement of non-flatness*

The output from a cold-rolling mill does not emerge loosely, as shown in Fig. 10.13, but instead is taken for coiling under tension, as shown in Fig. 10.15. The tension distribution across the strip width, indicated schematically in the figure, can then

be taken as a measure of non-flatness (i.e. a uniform tension distribution implies a flat strip). The measurement problem fell into two regions.

(*a*) Developing a reliable method for measuring on-line tension distribution across the width of a moving strip.
(*b*) Relating tension distribution quantitatively to flatness (as understood by the customer for the strip) and to control of flatness (as understood by a control engineer).

Each problem required several man-years for its solution.

For (*a*) a number of experimental devices, scanning or multi-probe, magnetic or mechanical, were developed and evaluated. A small number of designs evolved

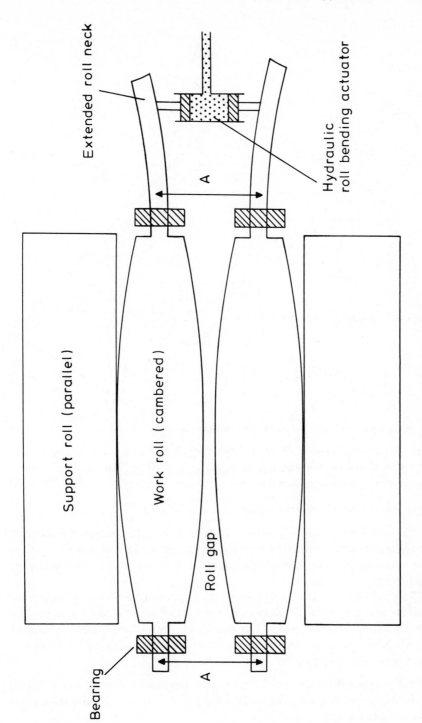

Fig. 10.14 *A strip-rolling mill illustrating provision for thickness and flatness control*

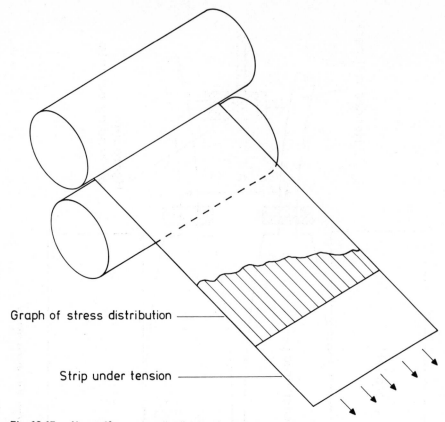

Graph of stress distribution ————————

Strip under tension ————————

Fig. 10.15   *Non-uniform stress distribution during strip-rolling*

until they became commercially available devices. One successful flatness measuring device works on the principle sketched in Fig. 10.16. Tension distribution is measured by a multi-segment roller, over which the strip passes.

Problem (*b*) has a number of aspects:

(*a*) Ensuring that spurious effects do not invalidate the measurement. For instance, non-uniform temperature distribution across the strip or slight mechanical misalignment might affect tension distribution, although they are not factors affecting the strip flatness.
(*b*) Representation of the tension distribution function by a set of numerical coefficients that are useful simultaneously for control purposes and for meaningful classification of flatness defects. This section of work required many laboratory tests on pieces of strip, followed by liaison with works quality-control inspectors from several different factories.

Finally a quantification of non-flatness by a low-order polynomial (Fig. 10.17) was found to be adequate. The coefficients y_1, y_2 and y_3 were found by curve fitting techniques.

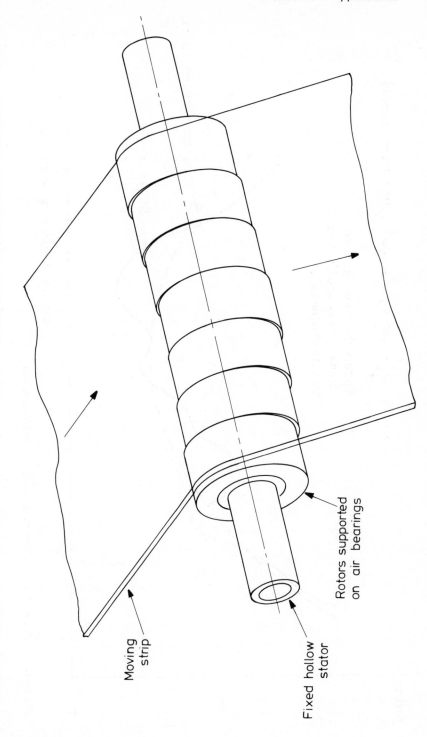

Moving
strip

Fixed hollow
stator

Rotors supported
on air bearings

Fig. 10.16   *A strip flatness meter – principle of operation of the Loewy-Robertson device*

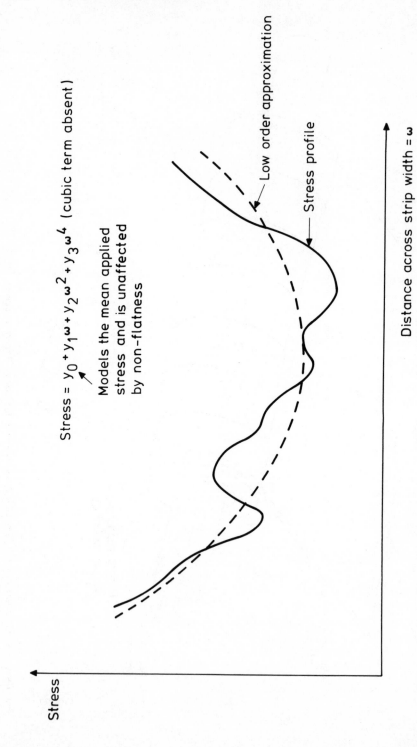

Stress = $y_0 + y_1 \omega + y_2 \omega^2 + y_3 \omega^4$ (cubic term absent)

Models the mean applied stress and is unaffected by non-flatness

Low order approximation

Stress profile

Distance across strip width = 3

Stress

Fig. 10.17   *The quantification of non-flatness by a low-order polynomial*

10.10.4 The relation between control variables and the non-flatness coefficients y_1, y_2 and y_3

Flatness correction is essentially achieved by making small changes to the operating conditions. Thus a linear small-perturbation model is adequate for control purposes. The model is of the form:

$$
\begin{bmatrix} \dot{x}_1 \\ \cdot \\ \cdot \\ \cdot \\ \dot{x}_4 \end{bmatrix} = A \begin{bmatrix} x_1 \\ \cdot \\ \cdot \\ \cdot \\ x_4 \end{bmatrix} + B \begin{bmatrix} u_1 \\ u_2 \end{bmatrix},
$$

$$
\begin{bmatrix} y_1 \\ \cdot \\ \cdot \\ \cdot \\ y_3 \end{bmatrix} = C \begin{bmatrix} x_1 \\ \cdot \\ \cdot \\ \cdot \\ x_4 \end{bmatrix}.
$$

The x vector represents the state of the two actuators that affect flatness. The first equation models the actuator dynamics. These were found by a mixture of methods described in Chapter 6, including the use of a correlator on-line.

The second equation represents the (assumed non-dynamic) relation between the state of the mill and the flatness of the strip produced. The matrix C is made up of sensitivity coefficients $\partial y_i / \partial x_j$. They were found from tests on the plant and from theoretical studies.

The actuator positions are:

$x_1 =$ degree of linear asymmetry (wedge) imparted to the roll gap by lifting one side and lowering the other,

$x_3 =$ degree of roll-bending imparted by hydraulic roll-bending cylinders.

(x_2 and x_4 are the time derivatives of the above variables.)

10.10.5 Design of a flatness control system

A flatness control system can be designed using the information derived earlier. However, it has to be taken into account that, when this is implemented, it may interact deleteriously with the already existing control loops. In this case, the existing control loop for thickness of the strip would interact strongly with any proposed flatness control scheme. Accordingly, an integrated scheme for control of both flatness and thickness is required.

If the mean gap between the unloaded rolls is designated x_5 and the mean thickness for the strip produced is designated y_4, then a model for both flatness and thickness can be written

$$
\begin{bmatrix} \dot{x}_1 \\ \cdot \\ \cdot \\ \cdot \\ \dot{x}_6 \end{bmatrix} = A \begin{bmatrix} x_1 \\ \cdot \\ \cdot \\ \cdot \\ x_6 \end{bmatrix} + B \begin{bmatrix} u_1 \\ \cdot \\ \cdot \\ u_3 \end{bmatrix},
$$

$$
\begin{bmatrix} y_1 \\ \cdot \\ \cdot \\ y_4 \end{bmatrix} = C \begin{bmatrix} x_1 \\ \cdot \\ \cdot \\ x_6 \end{bmatrix}.
$$

To illustrate the design of a controller for this example, we reduce the problem at this point so that only strip thickness and the parabolic coefficient of the flatness polynomial remain. These are the two most significant interacting variables and in this reduced form the problem is easily understood.

Strip thickness is designated y_1.

Strip flatness (parabolic coefficient) is designated y_2.

Thickness is controlled by simultaneous movement of the two dimensions marked A, in Fig. 10.14, and flatness is controlled by the hydraulic actuator operating on the extended roll necks. The signals applied to the two sets of actuators are designated u_1 and u_2.

However, the thickness control loop affects flatness since any attempt to change the roll gap produces unwanted roll bending. Similarly, the hydraulic actuator produces unwanted changes in the mean roll gap. Thus, the control loops interact.

To allow the use of a multivariable design package, a transfer function matrix G is set up as in Chapter 8, such that

$$
y(s) = G(s)u(s).
$$

For a particular mill, the matrix G has the numerical values

$$
G(s) = \begin{bmatrix} \dfrac{6}{s(1 + 0\cdot4s)} & \dfrac{-60}{s(1 + 0\cdot05s)} \\[3mm] \dfrac{-3}{s(1 + 0\cdot4s)} & \dfrac{160}{s(1 + 0\cdot05s)} \end{bmatrix}
$$

These values have been used as a multivariable design exercise in Section 8.4. An integrated control scheme for flatness and thickness for this (simplified) application then results.

Further information relevant to this application can be found in Leigh (1977*b*).

10.11 Control of a three-stand strip-rolling mill

Strip velocities and strip thicknesses interact strongly in this example. The relations have been heavily idealised and the dimensionality reduced to make the analysis tractable while retaining the main features.

Figure 10.18 shows the arrangement. The strip passes sequentially through the three rolling stands being successively reduced in thickness at each stand. The tensioning arms between the stands allow for small transient velocity mismatch.

The variables marked $y_1, \ldots, y_8$ are defined as below:

y_1 (strip thickness from stand 1),
y_2 (strip velocity from stand 1),
y_3 (length of strip between stands 1 and 2),
y_4 (strip thickness from stand 2),
y_5 (strip velocity from stand 2),
y_6 (length of strip between stands 2 and 3),
y_7 (strip thickness from stand 3),
y_8 (strip velocity from stand 3).

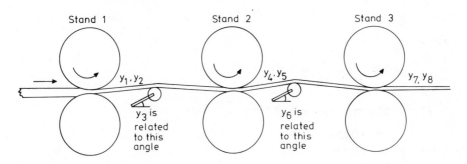

Fig. 10.18   *A three-stand tandem-rolling mill*

The variables defined $y_1, \ldots, y_8$ are to be controlled using variables defined $u_1, u_2, u_4, u_5, u_7, u_8$ as follows:

u_1 is to control y_1,
u_2 is to control y_3,
u_4 is to control y_4,
u_5 is to control y_2, y_5, y_8 together (i.e. to alter the overall speed of the mill),
u_7 is to control y_7,
u_8 is to control y_6.

(These control connections are those used in a normal mill.)

Considering only first-order relations and replacing the nonlinearities by linear approximations leads to the following transfer matrix $G(s)$ relating the u and y variables:

$$\begin{bmatrix}
\dfrac{6}{s(1+0.4s)} & 0 & 0 & 0 & 0 & 0 \\[3mm]
0 & g(s) & 0 & 0 & 0 & 0 \\[3mm]
\dfrac{0.9}{s^2(1+0.4s)} & \dfrac{g(s)}{s} & \dfrac{-0.56}{s^2(1+0.4s)} & \dfrac{-0.8g(s)}{s} & 0 & 0 \\[3mm]
\dfrac{4e^{-T_1 s}}{s(1+0.4s)} & 0 & \dfrac{4}{s(1+0.4s)} & 0 & 0 & 0 \\[3mm]
0 & 0 & 0 & g(s) & 0 & 0 \\[3mm]
0 & 0 & \dfrac{0.68}{s^2(1+0.4s)} & \dfrac{g(s)}{s} & \dfrac{-0.5}{s^2(1+0.4s)} & \dfrac{-0.85g(s)}{s} \\[3mm]
0 & 0 & 0 & \dfrac{3e^{-T_2 s}}{s(1+0.4s)} & \dfrac{2}{s(1+0.4s)} & 0 \\[3mm]
0 & 0 & 0 & 0 & 0 & g(s)
\end{bmatrix}$$

where

$$g(s) = \frac{1}{(1+2s)(1+s)}.$$

10.11.1 Typical numerical values

$y_1 = 0.002\,\text{m}$
$y_2 = 5\,\text{m s}^{-1}$
$y_3 = 4\,\text{m}$
$y_4 = 0.0015\,\text{m}$
$y_5 = 7\,\text{m s}^{-1}$
$y_6 = 4\,\text{m}$
$y_7 = 0.001\,\text{m}$
$y_8 = 11\,\text{m s}^{-1}$
$T_1 = 0.8\,\text{s}$
$T_2 = 0.57\,\text{s}$

10.11.2 Modelling the transport delays

The strip from stand 1 arrives at stand 2 after a time $T_1 = y_3/y_2$. The strip from stand 2 arrives at stand 3 after a time $T_2 = y_6/y_5$.

For simplicity, these delays can be regarded as constant, allowing analysis (although for investigation by simulation they will be allowed to vary).

From consideration of Fig. 10.18, interaction between control loops can be expected (and the presence of off-diagonal terms in the G matrix confirms this). Suppose, for instance, that control u_1 is used to reduce thickness y_1. To preserve mass flow, the strip velocity y_2 will automatically increase. The new thickness will arrive at stand 2 after a transport delay T_1 and will alter the outgoing velocity y_5. Ultimately, all the variables will be changed. The whole arrangement forms a challenging problem for integrated control system design using the methods of Sections 8.1–8.4.

10.11.3 Derivation of the G matrix

The three roll speeds are governed by a transfer function $g(s) = 1/(s + 1)(s + 2)$ and it is assumed that the outgoing strip velocity from each pair of rolls is equal to the peripheral speed of the rolls.

The strip thickness from the first roll gap is assumed to depend only on the roll gap setting (i.e. no changes in incoming strip conditions are allowed for). The roll gap can be altered through a second-order relation and when this is combined with rolling equation effects there results a transfer function $6/s(1 + 0.4s)$. The strip is subject to a transport delay T_1 before it reaches stand 2. At this stand the outgoing thickness depends on both the incoming thickness and on the roll gap transfer function. Similar relations affect the strip thickness from stand 3.

The strip between stands passes over a roller on an arm driven upwards by a torque motor. The arms must be maintained at particular angles – if the arms are pulled flat the strip will come under high tension and if the arms become too high the strip will loop over itself. The length of strip between stands is given by a relation of the form:

$$\text{length} = \text{initial length} + \int (\text{velocity mismatch between stands}) \, dt.$$

The outgoing velocity from each stand is given by a relation of the form (based on conservation of volume flow with invariant width):

$$\text{outgoing velocity} = \frac{\text{incoming velocity} \times \text{incoming thickness}}{\text{outgoing thickness}}.$$

Outgoing thickness depends on incoming thickness and on roll gap dimension.

Referring to the y variables defined, the equation for strip length y_3 between stands 1 and 2 becomes

$$y_3(t) = y_3(0) + \int_0^t (y_2 - y_5 y_4 / y_1) \, dt.$$

Differentiating and linearising leads to

$$\dot{y}_3(t) = y_2 - c_1 y_5 - c_2 y_4 + c_3 y_1,$$

where the c are constants.

A similar relation is defined for the strip length between stands 2 and 3.

10.12 Integrated control of a petrochemical complex

At British Petroleum's Baglan Bay plant (South Wales, U.K.) an ethylene cracker is at the centre of the complex, providing ethylene, gasoline, fuel gas and by-products to a variety of other processes on the site. The plant consists of some eighteen interconnected processes that, from the main raw material, naphtha, produce thirteen different products including propylene, isopropyl alcohol, styrene, PVC and caustic soda.

This very large plant on a site of about 600 acres is under integrated control based on hierarchical principles at three levels. At the lowest (individual process) level, conventional control has been implemented. The next (supervisory) level is concerned with balancing operations of the overall plant by setting individual process production levels.

The highest level is concerned with optimisation and scheduling of the plant to integrate the total operation with raw material delivery, patterns and prices and with predicted demands for the wide variety of products.

The hardware configuration includes:

(*a*) An ICL central data processing computer (196 K store).
(*b*) A Ferranti Argus 500 message switching computer providing an information link between all control centres in the plant.
(*c*) Three Argus 500 plant computers.
(*d*) A Kent telemetry system.

In this scheme there has been an emphasis on economic factors from the start, with mathematical models being developed to estimate cost benefits and with economic data and targets being displayed routinely to plant line management.

10.13 Energy management in large buildings

Increasingly sophisticated control systems are coming into use for the control of temperature in large office blocks and similar buildings. Programs are able to take into account the utilisation pattern of the buildings as well as internal and external temperatures.

Special problems encountered include:

(*a*) the sudden unexpected large changes in outside temperature, and
(*b*) the fundamental changes in system duty between Summer and Winter so that a different *structure* of control system is required in the two instances.

System SDC 8003 of ITT offers a variable time control program that, using a learning function, ensures that internal temperatures are not reached too early nor maintained longer than is strictly necessary to satisfy the occupation pattern. A further program selects and mixes in optimum proportions the source air for air-conditioning, using a mixture of fresh air and recirculated air and taking into account humidity as well as temperature.

Another growing application area is to chains of supermarket stores where integrated energy policies covering freezer cabinets and store heating and lighting have been shown to be profitable.

10.14 Some general points concerning the control of batch processes

In the terminology of batch processes, control is often divided into so-called static and dynamic control.

If one imagines firing a missile at a target as being a batch process then static control consists of deciding on the weight, fuelling, orientation, general initial state and in-flight pre-programmed control actions. Dynamic control consists of those actions most usually associated with a 'guided missile', i.e. closed-loop modification to the trajectory in the light of measured information received.

Good static control is an essential pre-requisite of satisfactory batch process operations (but notice that such static control may be nothing more than ensuring accurate weighing of constituents — there may be no familiar control system adjuncts in sight). Dynamic control is necessary to compensate for process variations that cannot easily be taken into account in the static control laws.

Control engineers become accustomed to using linear models. However static control of batch processes will often involve the use of a comprehensive nonlinear process model. Naturally, therefore, it will often need the assistance of process experts to design the static control models.

10.14.1 Deciding on the necessary accuracy for which static control of the initial conditions should aim

Attaining high accuracy of a mixture of materials is very expensive when large quantities are being considered. It is important therefore not to specify unnecessarily high accuracies. Ideally, rather than a blanket statement on accuracy, each aspect should be considered individually and a 'necessary accuracy profile' produced. One method for doing this is illustrated in the idealised example below.

Let the initial weights of process ingredients be given as $x_i(0)$, $i = 1, \ldots, n$, with the x_i being individual elements. Let the (simplified) aim be to obtain correct composition of the element x_1 at the end, T_b, of the process, i.e. it is required that, with c given,

$$\frac{x_1(T_b)}{\Sigma_{i=1}^n x_i(T_b)} = c.$$

Assume next that the ideal static control can be achieved but that errors in weighing perturb the initial mix. A numerical sensitivity analysis is now undertaken with the aid of the static control model, which by implication exists (although an alternative is that an empirical static control procedure has been evolved without the use of an explicit model).

The sensitivity coefficients $\partial x_1(T_b)/\partial x_i(0)$, $i = 1, \ldots, n$, are determined by repeated computations. In a reasonably well-controlled situation, the expected perturbations in the x_i should be small enough not to invalidate the linearity assumption implicit in the sensitivity analysis.

Given the sensitivity coefficients, it is obvious how to use these in coming to an estimate of necessary accuracies for initial conditions. The easiest procedure is to take a 'worst case' approach and assume that errors can be added. A more scientific approach may be required in which case data on the statistical interdependence of the expected variations in the x_i will need to be obtained.

10.14.2 Batch-to-batch adaptation of the static control strategy

It will have occurred to the reader that static control as described above is open-loop control. Like all open-loop control, the results achieved depend on the accuracy of the model used. To bring in an element of feedback to compensate for model deficiencies and in particular to compensate for slow changes in the process, batch-to-batch adaptation can be practised. Let the process variables be represented by vector x driven from $x(0)$ to $x(t_f)$. Suppose that the desired value for $x(t_f)$ is the vector x_d. Then an error vector $e = x_d - x(t_f)$ can be used to drive batch-to-batch adaptation in an obvious way.

Considering only the error e_i in the ith variable and given a knowledge of the sensitivity coefficients $\partial e_i/\partial x_j(0)$, $j = 1, \ldots, n$, as well as $\partial e_i/\partial t_f$ and further sensitivity coefficients connecting e_i with the stored control strategy, then provided that each process batch takes place under identical conditions, there are a number of ways in which e_i could be brought to zero. Clearly, when considering the whole vector e, some form of linear transformation from e to the corresponding corrections can easily be devised. However, all adaptive loops can pose stability problems under adverse conditions and for this reason an attenuation factor will usually be introduced so that only a proportion of the error is corrected at each updating operation.

10.14.3 A more sophisticated method for batch-to-batch adaptation

The adaptation described in Section 10.14.2 uses as the current control strategy, that determined as being correct for the previous batch. The procedure described in this section is an improvement in that it identifies the current batch and applies any modifications to the control strategy immediately.

Batch n is controlled initially using the strategy that was valid for the previous batch (batch $n - 1$). Data logging is undertaken until sufficient data have been obtained to allow reliable identification of the characteristics of the current batch. The actual method of identification is not important.

Once the current batch has been identified, a new control strategy must be calculated and implemented for the time remaining in the batch.

10.14.4 Linearising a batch process about the trajectory for which the static control system implicitly aims

The emphasis here is not on the technique of linearisation but on the application.

Many practising engineers seem to think that linearisation involves a *fixed* operating point. However, a nonlinear batch process will often range over a large operating region. Dynamic control can consist in operating on the deviations from an implicit process trajectory that is contained in and aimed for by the static control system. Notice, however, that if a nonlinear system is linearised about a variable trajectory, then the time-invariant nonlinear model will be found to have been exchanged for a time-varying linear model, as the following example makes clear.

Let the nonlinear process model be of the form

$$\dot{x} = f(x, u), \quad x(0), u_s(t), t \in [0, T_b] \text{ given,}$$

and let the scheduled control $u_s(t)$ produce the statically-desired trajectory $x_s(t)$, $t \in [0, T_b]$ with $x_s(0) = x(0)$.

The linearised model is

$$\delta x(t) = \left.\frac{\partial f}{\partial x}\right|_{x=x_s(t)} \delta x(t) + \left.\frac{\partial f}{\partial u}\right|_{u=u_s(t)} \delta u(t).$$

Clearly the coefficients $\partial f/\partial x$, $\partial f/\partial u$ change in a known manner as the process progresses. A practical way to use this approach is to split the batch process time T_b into a few regimes and within each regime to regard the coefficients $\partial f/\partial x$, $\partial f/\partial u$ as constant at reasonable mean values. The number of regimes should be chosen so that the performance obtained is not significantly worse than if the continuously linearised model had been implemented.

Chapter 11

DC motors and servomechanisms

11.1 DC motors in control loops

DC motors remain very important for control system applications in two main areas:

(*a*) High-power drives where precise variable-speed control is required.
(*b*) Servomechanisms.

Although AC machines, in conjunction with solid-sate electronics have made great progress, DC machines have also been the subject of improvement. For small power applications, permanent magnet motors are now very competitive. Compared with their wound-field counterparts they have a higher torque at zero speed (since armature reaction has little effect because of the low permeability of modern ceramic permanent magnets) and an almost linear torque–speed characteristic. For very low-power applications, small moving-coil DC motors can complete thousands of stop–start cycles per second. They find most application in computer peripherals, such as tape readers.

11.1.1 Motor speed control through DC machines
The most simple speed control system is unidirectional, essentially operating in only one quadrant of the speed–torque graph. In this type of system, if a lower speed is demanded, the motor coasts to its new value, there being no provision for torque reversal. A refinement, in the form of dynamic braking, allows improvement in this respect, but a system operating fully in all four quadrants of the speed–torque graph would generally be defined to a be a servomechanism rather than a speed control system.

A DC motor supplied by a well-designed solid-state source and controlled in open loop (i.e. with no measurement of speed) can achieve a wide range of speed variation (20:1 ratio) with a fall in speed between no load and full load of about $0.05\omega_{max}$ where ω_{max} is the highest speed within the controlled range. If a better specification is needed or if the transient behaviour is critical, a closed-loop system will be needed.

For large systems, the Ward–Leonard system was much favoured in the past. It basically consists of a constant-speed motor driving a generator. The generator provides a variable voltage to the armature of the motor whose speed is to be controlled by varying the generator excitation. Diesel-electric locomotives use this principle. The diesel engine drives a generator that supplies a number of driving motors slung directly on the locomotive axles.

For industrial speed control applications, thyristor drives have replaced Ward–Leonard systems.

Latest speed control systems have desired speed set by crystal controlled oscillator and shaft speed measured by digital tachogenerator. Steady-state accuracies can approach 0·0001% with time constants down to a few milliseconds. A very high mechanical stiffness is of course needed in the drive train if such values are to be attained.

11.1.2 The transfer function of a DC motor with constant field

Define armature e.m.f., voltage, current, resistance and inductance by E_a, V_a, i_a, R_a and L_a. Shaft angular velocity, effective inertia and viscous friction coefficient by ω, J, D. The constant friction torque, the driving torque and the load torque by T_f, T_g and T_L. Constants k_E and k_T are defined by the relations

$$E_a = k_E\omega, \quad T_g = k_T i_a.$$

The machine equations are then, neglecting shaft torsion,

$$V_a = L_a \dot{i}_a + i_a R_a + k_E\omega,$$

$$T_g = J\dot{\omega} + D\omega,$$

combining and Laplace transforming leads to the transfer function

$$G(s) = \frac{\omega(s)}{V(s)} = \frac{k_T}{(sL_a + R_a)(sJ + D) + k_E k_T},$$

$$= \frac{1}{k_E(1 + s\tau_1)(1 + s\tau_2)},$$

where $\tau_1 = -1/\alpha_1$, $\tau_2 = -1/\alpha_2$ and where α_1 and α_2 are the roots of the equation.

$$s^2 L_a J + s(L_a D + R_a J) + R_a D + k_E k_T = 0.$$

In many practical situations, D and L_a are very small and can be neglected, then

$$G(s) = \frac{1}{k_E(1 + s\tau_m)(1 + s\tau_E)},$$

where $\tau_m = R_a J/k_E k_T$ (the mechanical time constant), and $\tau_E = L_a/R_a$ (the electrical time constant).

In a high-performance servomechanism, the destabilising influence of shaft compliance should be examined, a higher-order model representing this effect being set up and analysed for this purpose. Although compliance in the main shaft between motor and load is the most likely source, destabilising influences can also arise from the motor–tachometer shaft.

11.1.3 Measurement of motor parameters under assumptions of linearity

The constant k_T can be measured from a test conducted at constant speed, measuring torque by dynamometer and plotting this against armature current, then

$$k_T = \frac{dT}{di_a}.$$

The constant k_E can be measured by using the motor as a generator and measuring no-load armature e.m.f. against armature angular velocity

$$k_E = \frac{dE_a}{d\omega}.$$

The electrical time constant $\tau_E = L_a/R_a$ can be measured by applying a step voltage and estimating the constant graphically from the current *versus* time graph. The mechanical time constant $\tau_m = R_a J/k_E k_T$ can be estimated from the step response of the angular velocity against time provided that the electrical time constant does not interact too much in the measurement. Alternatively τ_m can be estimated from a frequency response test since it can be identified with a break point. Again, if the electrical time constant is similar to the mechanical time constant, the determination is not so simple.

11.1.4 Measurement of motor characteristics where the nonlinearities in load torque and in field inductance cannot be ignored

Where a motor operates over a wide speed range and has a variable field voltage applied, linearity assumptions fail. Load torque must now be considered a function of angular velocity and field inductance, a function of field current.

Both functions can be determined by simple tests as follows.

11.1.5 DC machines — tests to determine inertia, nonlinear load characteristic and nonlinear inductance of the field winding

11.1.5.1 Determination of the inertia of motor and load: The motor (with load) is run at a steady speed ω_0 and the electrical inputs V_0 and I_0 are measured. The electrical power is switched off and a tachometer record is made of the angular velocity running down to zero. Let T_0 be the mechanical retarding torque at angular velocity ω_0. Then during steady running

$$\omega_0 T_0 = I_0 V_0 - \text{electrical losses}.$$

(The electrical losses are calculable to a reasonable accuracy, so that T_0 can be assumed known.)

At the instant when the supply is switched off, the instantaneous rate of change of energy is $\omega_0 T_0$. The energy stored in the rotating shaft is $\frac{1}{2}J\omega^2$.

Thus

$$\frac{d}{dt} \tfrac{1}{2}J\omega^2 \bigg|_{\omega_0} = \omega_0 T_0.$$

(Here we have assumed that the inertia J is invariant with speed – usually a reasonable assumption.)

This leads to

$$J = T_0/\dot{\omega}|_{\omega=\omega_0}.$$

$\dot{\omega}|_{\omega=\omega_0}$ is easy to pick off from the tachometer record so that J can be calculated. (I have carried out tests on a motor driving a large disc where the inertia was easily calculable and found that good accuracy was obtainable.)

11.1.5.2 Determination of the load characteristic: Further useful information on the load characteristic can now be extracted from the record of angular velocity against time.

At any angular velocity ω_1

$$J = T_1/\dot{\omega}|_{\omega=\omega_1}.$$

Now that J is known, $\dot{\omega}|_{\omega=\omega_1}$ is picked off the tachometer run-down graph and T_1 can be determined. In this way the complete load characteristic can be determined from a single run-down test.

Some practical points: We have found that some mechanisms have different load characteristics when turned in the opposite direction; any mechanism involving a large gear train will tend to have a temperature sensitive load characteristic. (This temperature will often be influenced more by the intensity of the past work load than by the ambient temperature.) These temperature effects are large and cannot be ignored. A little practice in interpreting angular velocity run-down graphs soon results in insight. For instance, a graph such as Fig. 11.1 indicates a significant static friction in the load.

11.1.5.3 Determination of the field inductance: The maximum working-field voltage is applied instantaneously to the field and the current rise recorded on an ultraviolet recorder or other suitable device.

We indicate that the inductance L may be a function of current i by the notation $L(i)$. Then if V is the applied voltage and R is the field resistance

$$V = iR + L(i)\frac{di}{dt}.$$

At any particular current i_1,

$$L(i_1) = \frac{V - i_1 R}{di/dt|_{i=i_1}}.$$

And by a straightforward graphical technique the inductance as a function of current can be determined and (say) stored in a table for use in simulation. Note that allowing for the nonlinearity (caused by saturation of the iron in the machine) is not an academic exercise – the field inductance will usually fall to less than half its zero current value before maximum field current is reached.

One word of warning. If, as is likely, tests are being conducted with an *ad hoc* assemblage of temporary equipment, care must be taken that the test current to the field can be switched off. A switch able to break a DC current of the magnitude expected must be included for safety purposes.

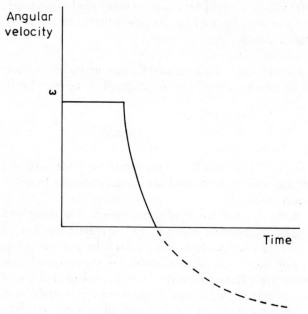

Fig. 11.1   *A run-down test on a mechanism having static friction*

11.2 Servomechanisms

Servomechanisms are designed for applications using straightforward methods [see for instance Wilson (1970)] and little more needs to be said here. However, design of systems for incremental motion seems to be a neglected area in the literature, so we devote a little space to that topic.

Finally, we illustrate how servomechanisms are typically embedded within larger industrial control systems where there are features that require special design attention.

11.2.1 Incremental motion
Digitally controlled systems often require incremental motion of actuators. Two useful design guidelines are given below.

11.2.2 Optimal incremental motion
Consider the case where a load of inertia J_L has to be moved from rest through an angle θ in a time t_c against the resisting torque T_L. It is assumed that an armature-

controlled motor of torque constant k_T, armature resistance R_a and moment of inertia J_m is to be used. If the aim is to achieve this movement with minimum energy dissipation where the energy is defined by

$$\int_0^{t_c} i_a^2(t)\, dt,$$

then the optimal velocity profile is parabolic, given by

$$\omega(t) = 6\theta \left(\frac{t_c - t}{t_c^3} \right) t.$$

In practice, this parabolic profile is approximated by a triangular or trapezoidal profile. If the efficiency with the (optimal) parabolic profile is considered to be 100%, then it is found that the triangular profile has an efficiency of 75% with the trapezoidal profile being almost 90% efficient. For more details see Electro-craft Corporation (1975).

11.2.3 Optimal gear ratio for incremental movement

It can be shown that, for movement over the angular increment θ_L with minimum energy, the motor of efficiency η should be coupled to the load by a gearbox of ratio (suffixes L, m indicate load and motor quantities respectively):

$$N = \left\{ \frac{J_L}{J_m} \left[1 + \frac{\eta}{12} \left(\frac{T_L}{\theta_L} \frac{t_c^2}{J_L} \right)^2 \right]^{1/2} \right\}^{1/2}.$$

Only in the case where $T_L = 0$ do we obtain the familiar relation

$$N = \left(\frac{J_L}{J_m} \right)^{1/2}.$$

The graph of efficiency against gear ratio is rather flat near the optimum, and if the gear ratio is incorrect by 10% the efficiency falls by only about 1%.

11.2.4 Stepper motors as incremental positioning actuators

A stepper motor moves one known angular increment each time it receives a pulse. Having stopped, it has an inherent resistance to displacement from its position. For position control, it is only necessary to send out a discrete number of pulses to the motor and the load shaft will move precisely as required. All that is necessary is to ensure that the pulse rate and the inertia of the load are within the motor specification. This is an open-loop position control system.

By attaching a shaft encoder to the stepper motor rotor and using additional feedback logic, the performance can be improved and in particular the degree of damping can be controlled.

For some applications, a stepper motor can be driven directly from the normal AC supply to provide a high-torque low-speed (50 r.p.m.) drive suitable for actuation.

Stepper motors and techniques for their application are described in, for instance, the Sigma Stepping Motor Handbook (1972).

11.2.5 Cutting steel bar to length while it is moving at high speed

Steel bar is produced in long sections by hot rolling. It is cut into lengths while still travelling at high speed. The control aim is for the steel bar to be cut into lengths within customer tolerances so that no odd unsaleable lengths are produced at the end of the run. This requires prediction of the total length of the batch, precise measurement of length (by optical means) and control of the actual cutting operation. Here we consider only the cutting operation which is performed on the moving bar by a so-called 'flying shear'. This consists of two discs mounted above and below the bar to be cut. The cutting blades are set into the perimeter of the discs. The control problem is to accelerate the discs so that, when the blades begin to cut the bar, their linear velocity matches that of the moving bar. Once the bar is cut, the discs must be decelerated and made ready for the next cut. The bar will be moving at a speed in excess of 20 metres/sec and the cut must be made to an accuracy of approximately one centimetre.

The accuracy of cutting necessarily depends on the repeatability of acceleration and the meeting of the specification requires careful control design.

11.2.6 Cutting plate glass to length

The cutting to length of a moving ribbon of plate glass also requires speed matching, but here the speed of the ribbon is slow and a perpendicular cross cut can be achieved by using two servomechanisms. One servomechanism moves the cutting head lengthwise in synchronism with moving glass. The second moves the head across the glass. The cutting head thus moves diagonally with respect to fixed axes to achieve a perpendicular cut across the glass.

References

Auslander, D. M., Takahashi, Y., and Tomizuka, M. (1978) 'Direct digital control: practice and algorithms for microprocessor application', *Proc. IEEE* **66**, No. 2, pp. 199–208.

Bristol, E. H. (1966) 'On a new measure of interaction for multivariable process control', *IEEE Trans. Automatic Control* **AC-11**, No. 1, pp. 133–134.

Buschart, R. J. and Hohlfeld, E. F. (1978) 'Users' experience with programmable controllers', *Instrum. Technol.* **25**, No. 1, January, pp. 55–59.

Chiu, K. C., Corripio, A. B., and Smith, C. L. (1973) 'Digital control algorithms', *Instrum. Contr. Syst.*, October, pp. 57–59, November, pp. 55–58, December, pp. 41–43.

Clarke, D. W. and Gawthrop, P. J. (1979) 'Self-tuning control', *Proc. IEE* **126**, No. 6, pp. 633–640.

Dahlin, E. B. (1968) 'Designing and tuning digital controllers', *Instrum. Contr. Syst.* **41**, No. 6, pp. 77–83 and pp. 87–91 (in two parts).

Dale-Harris, L. (1961) *Introduction to Feedback Systems*, John Wiley and Sons, New York.

Davies, M. S. (1978), 'Simpler filter for fast faithful feedback', *Pulp and Pap. Can. (Canada)* **79**, No. 5, p. 108.

Davis, T. P. and Smith, C. A. (1977) 'Evaluation of various control strategies for a power plant fuel oil heating control system', see Instrument Society of America (1977), pp. 47–55.

Desai, v. K. and Fairman, F. W. (1971) 'On determining the order of a linear system', *Mathematical Biosciences* **12**, pp. 217–224.

Donoghue, J. F. (1977) 'Review of control design approaches for transport delay processes', *ISA Trans.* **16**, No. 2, pp. 27–35.

Electro-craft Corporation (1975) *DC Motors, Speed Controls, Servosystems: An Engineering Handbook*, Pergamon Press, Oxford.

Fisher, D. G. and Seborg, D. E. (1976) *Multivariable Computer Control, A Case Study*, North Holland Publishing Co., New York.

Funk, G. L. and Smith, D. E. (1974) 'Estimating economic incentives for computer control systems – an applications approach', *Trans. IEE Ind. Appl. USA* **IA-15**, No. 4, pp. 394–398.

Goff, K. W. (1966) 'Dynamics in direct digital control – I, II', *J. Inst. Soc. of America*, November, pp. 45–49, December, pp. 44–54.

Gordon, D. and Spencer, R. D. (1977) 'Economic aspects of distributed systems', *International Conference on Distributed Computer Control*, 26–28 September 1977, Birmingham, England, publ. IEE London, England, pp. 131–133.

Grensted, P. E. W. (1962) 'Frequency response methods applied to nonlinear systems' in the book *Progress in Control Engineering – 1*, Heywood and Co. Ltd., London.

Hickey, J. (1978) 'A look at programmable controllers today', *Instrumentation and Control Systems* **51**, No. 5, pp. 23–31.

Instrument Society of America (1977), Proceedings of the 1977 Spring industry oriented conference 'Instrumentation in the Chemical and Petroleum Industries', Anaheim, California.

Jacobs, O. L. R., Hewkin, P. F., and While, C. (1980) 'On-line computer control of pH in an industrial process', *Proc. IEE* 127, Pt. D, No. 4, pp. 161–168.

Kallina, G. (1981) 'Nonlinear problems in complex control systems', *Process Automation* 1, pp. 44–49.

Kalman, R. E. (1960), 'A new approach to linear filtering and prediction problems', *Trans. ASME, J. Basic Eng.* 82, pp. 35–45.

Kalman, R. E. and Bertram, J. E. (1960) 'Control system analysis and design via the second method of Lyapunov', *Trans. ASME* 82, Series D, pp. 371–400.

Kalman, R. E. and Bucy, R. S. (1961), 'New results in linear filtering and prediction theory', *Trans. ASME, J. Basic Eng.* 83, pp. 95–108.

Kochhar, A. K. and Parnaby, J. (1978), 'Comparison of stochastic identification techniques for dynamic modelling of plastics extrusion processes', *The Institution of Mechanical Engineers*, proceedings, 192, No. 28, pp. 299–309.

Leigh, J. R. (1977a) 'A four term controller for precise control to a temperature curve', *Instrumentation and Control*, June, pp. 33–35.

Leigh, J. R. (1977b) 'Control systems for strip rolling mills', *Measurement and Control, U.K.* 10, pp. 433–437.

Leigh, J. R. (1979) 'Bridging the gap between Measurement and Control', *Journal of the Institute of Measurement and Control* 12, No. 1, pp. 26–28.

Leigh, J. R. (1980) 'Modelling principles and simulation' in the book *Modelling of Dynamic Systems*, Vol. I (Edited by H. Nicholson), Peregrinus, Stevenage, U.K.

Leigh, J. R. and Williams, R. V. (1972) 'Developments in the control of iron and steelmaking processes' in the book *Advances in Extractive Metallurgy and Refining*, published by the Institute of Mining and Metallurgy, U.K.

Litchfield, R. J., Campbell, K. S., and Locke, A. (1979) 'The application of several Kalman filters to the control of a real chemical reactor', *Trans. Inst. Chem. Eng. (GB)* 57, No. 2, pp. 113–120.

Lopez, A. M., Murill, P. W., and Smith, C. L. (1969) 'Tuning PI and PID digital controllers', *Instruments and Control Systems* 42, No. 2, pp. 89–95.

Marshall, J. E. (1979) *The Control of Time Delay Systems*, Peregrinus, Stevenage, U.K.

Martin, J. (Jr.), Corripio, A. B., and Smith, C. L. (1977) 'Comparison of tuning methods for temperature control of a chemical reactor', see Instrument Society of America (1977), pp. 31–35.

McRuer, D. (1980) 'Human dynamics in man–machine systems', *Automatica* 16, No. 3, pp. 237–254.

Munro, N. and Ibrahim, T. A. S. (1975) 'Design of sampled data multivariable systems using the inverse Nyquist array', *Int. J. Control* 22, No. 3, pp. 297–311.

Nicholson, H. (ed.) (1980, 1981) *Modelling of Dynamical Systems*, Vols. I & II, Peregrinus, Stevenage, U.K.

Page, M. (1979) 'Microprocessor implementation of the Kalman filter', *Microelectron. J.* 10, No. 3, pp. 16–22.

Richalet, J., Rault, A., Testud, J. L., and Papon, J. (1977) 'Model algorithmic control of industrial processes', paper c-10, Proceedings of the 4th IFAC Symposium, Tbilisi, USSR, published by North-Holland Publishing Company (Edited by Van Nauta Lemka).

Richalet, J., Rault, A., Testud, J. L., and Papon, J. (1978) 'Model predictive heuristic control: Applications to industrial processes', *Automatica* 14, pp. 413–428.

Rose, E. and Radmanesh, A. (1981) 'Feedforward control of a sinter strand process', Proc. IEE Conf. Control and its Applications, Warwick, U.K., 23–25, 1981, pp. 340–345.

Rosenbrock, H. H. (1969) 'Design of multivariable control systems using the inverse Nyquist array', *Proc. IEE* 116, No. 11, pp. 1929–1936.

Rosenbrock, H. H. (1974) *Computer Aided Control System Design*, Academic Press, London.

Ross, C. W. (1977) 'Evaluation of controllers for dead-time processes', see Instrument Society of America (1977), pp. 87–97.

Rovira, A. A., Murrill, P. W., and Smith, C. L. (1969) 'Tuning controllers for setpoint changes', *Instruments and Control Systems* 42, No. 12, p. 67.

Sayers, B. and Moore, C. F. (1977) 'Application of a simple discrete multivariable strategy to a laboratory scale autoclave', see Instrument Society of America (1977), pp. 79–85.

Shinskey, F. G. (1979) *Process Control Systems*, McGraw-Hill Book Co., New York.

Sigma Instruments Inc. (1972) *Stepping Motor Handbook*.

Smith, C. A. and Groves, F. R. (1977) 'Empirical second order nonlinear process model development and application', see Instrument Society of America (1977), pp. 37–46.

Smith, C. L. (1972) *Digital Computer Process Control*, Intex International Publishers, London.

Smith, H. W. (1977) 'Computer control justifies its costs', *Com. Controls and Instrum. (Canada)* **16**, No. 8, pp. 28–30.

Smith, O. J. M. (1959) 'A controller to overcome dead-time', *ISAJ* **6**, No. 2, pp. 28–33.

Stout, Th. M. (1973) 'Economic justification of computer control systems', *Automatica* **9**, pp. 9–19.

Takahashi, Y., Tomizuka, M., and Auslander, D. M. (1975) 'Simple discrete control of industrial processes', *Transactions of the ASME, Journal of Dynamic Systems, Measurement and Control*, pp. 354–361.

Teodorescu, D. (1973) *Entwurf nichtlinearer Regelsysteme mittels Abtastmatrizen*, A. Huthig, Heidelberg.

Tomizuka, M., Auslander, D. M., and Takahashi, Y. (1977) 'Simple finite time settling control and manipulated variable softening for reverse reaction, overshoot and oscillatory processes', Presented at the ASME Winter Annual meeting, November.

Unbehauen, H., Schmid, Chr., and Bottiger, E. (1976) 'Comparison and application of DDC algorithms for a heat exchanger', *Automatica* **12**, pp. 393–402.

Uronen, P. and Yliniemi, L. (1977) 'Experimental comparison and application of different DDC-algorithms', Paper M1-4, 5th IFAC/IFIP Conf. on Digital Computer Application to Process Control, North Holland, Amsterdam.

Wiberg, D. M. (1971) *State Space and Linear Systems*, Schaum Outline Series, McGraw-Hill Book Co., New York.

Wick, H. J. (1978) Estimation of the core temperature of ingots in a soaking pit with modified Kalman filter' (In German), *Gas Wärme Int.* **27**, No. 10, pp. 548–552.

Wilson, D. R. (ed.) (1970) *Modern Practice in Servomechanism Design*, Pergamon Press, Oxford, U.K.

Wonham, W. M. (1967) 'On pole assignment in multi-input controllable linear systems', *IEEE Trans. Automatic Control* **AC-12**, pp. 660–665.

Wood, B. I. (1973) 'Simulation and control of the blast furnace process', Proc. 5th UKAC Convention, University of Bath, U.K.

Ziegler, J. G. and Nichols, N. Z. (1942), 'Optimum settings for automatic controllers', *Trans. ASME* **64**, No. 11, p. 759.

Bibliography

The following conference proceedings contain many relevant papers.

IEE Conference publications
>No. 150 (1977) 'Displays for man–machine systems'
>No. 153 (1977) 'Distributed computer control systems'
>No. 161 (1978) 'Centralised control systems'
>No. 172 (1979) 'Trends in on-line computer control systems'
>No. 194 (1981) 'Control and its applications'

Purdue International Workshop on Industrial Computer Systems (1975), 'Guidelines for the design of man–machine interfaces for process control', Purdue University, Indiana, U.S.A.

Proceedings of the 7th Triennial World Congress of the International Federation of Automatic Control, Helsinki, Finland 12th–16th June 1978.

Proceedings of the 2nd IFAC/IFIP Conference on Software for Computer Control (edited by Novak, N.), Prague, Czechoslovakia 11–15 June 1979. Published by Pergamon Press.

Proceedings of the 6th IFAC/IFIP International Conference on Digital Computer Applications to Process Control, Düsseldorf, 1980.

Amerongen, Van. J. and Udink Ten Cate, A. J. (1975) 'Model reference adaptive autopilots for ships', *Automatica* 11, pp. 441–449.

Anderson, N. A. (1980) *Instrumentation for Process Measurement and Control*, Chilton Book Company, Radnor, Pennsylvania.

Aström, K. J. and Wittenmark, B. (1973) 'On self-tuning regulators', *Automatica* 9, pp. 185–199.

Aström, K. J., Borrison, U., Ljung, L., and Wittenmark, B. (1977) 'Theory and applications of self-tuning regulators', *Automatica* 13, pp. 457–476.

Bibby, K. S., Margulies, F., Rijnsdorp, J. E., and Withers, R. M. J. (1975) 'Man's role in control systems', Plenary Papers, 6th IFAC Congress, Boston, U.S.A.

Bierman, G. J. and Thornton, C. L. (1977) 'Numerical comparison of Kalman filter algorithms', *Automatica* 13, pp. 23–25.

Billings, S. A. and Harris, C. J. (1981) 'Self tuning and adaptive control: theory and applications', *IEE Control Engineering Series*, Peregrinus, Stevenage, U.K.

Blight, J. D. and McClamrock, N. H. (1977) 'Graphical stability criteria for nonlinear multi-loop systems', *Automatica* 13, pp. 189–190.

Borisson, U. and Syding, R. (1976) 'Self-tuning control of an ore crusher', *Automatica* 12, pp. 1–7.

Bosman, D. (1974) *A Design Methodology for Man–Machine Systems*, T.H. Twente, Netherlands, pp. 1–8.

Box, G. E. P. and Jenkins, G. M. (1970) *Time Series Analysis – Forecasting and Control*, Holden-Day, San Francisco.

Bristol, E. H. (1976) 'Designing and programming control algorithms for DDC systems', *Contr. Eng.* **24**, No. 1, pp. 24–26.

Bristol, E. H. (1977) 'Pattern recognition: an alternative to parameter identification in adaptive control', *Automatica* **13**, pp. 197–202.

Cameron, J. F. and Clayton, C. G. (1971) *Radioisotope Instruments*, Pergamon Press, Oxford, U.K.

Cegrell, T. and Hedqvist, T. (1975) 'Successful adaptive control of paper machines', *Automatica* **11**, pp. 53–59.

Chiu, K. C., Corripio, A. B., and Smith, C. L. (1972) 'Process models for controller tuning', *Inst. and Cont. Systems* **45**, No. 1, pp. 84–88.

Clarke, D. W., Cope, S. N., and Gawthrop, P. J. (1975) 'Feasibility study of the application of microprocessors to self-tuning regulators', OUEL Report, 1137/75.

Considine, D. M. (1974) *Process Instruments and Controls Handbook*, McGraw-Hill Book Company, New York.

Corripio, A. B. and Smith, C. L. (1971) 'Computer simulation to evaluate control strategies', *Instruments and Control Systems* **44**, No. 1, pp. 87–91.

Costello, R. G. and Higgins, T. J. (1966) 'An inclusive classified bibliography pertaining to modelling the human operator as an element in an automatic control system', *IEE Trans. Human Factors in Elect.* **7**, No. 4, pp. 174–181.

Courtiol, B. and Landau, I. D. (1975) 'High speed adaptation system for controlled electrical drives', *Automatica* **11**, pp. 119–127.

Cundall, C. M. (1972) 'A review of operator communication methods in computer control', Proceedings of 1972 IFAC 5th World Congress, Instrument Society of America, paper 5-4, pp. 1–11.

Davies, W. D. T. (1967) 'Control algorithms for DDC', *Instrument Practice* **21**, pp. 70–77.

Davison, E. J. and Goldenberg, A. (1975) 'Robust control of a general servomechanism problem: the servo compensator', *Automatica* **11**, pp. 461–471.

Davison, E. J., Taylor, P. A., and Wright, J. D. (1980) 'On the application of tuning regulators to control a commercial heat exchanger', *IEE Trans. Automatic Control* **AC25**, No. 3, pp. 361–374.

Dowson, M., Wilkinson, P. T., Collins, B., McBride, B., and Milne, R. (1979) 'The demos multiple processor', EURO IFIP 79, North-Holland Publishing Company, pp. 679–684.

Droffelaar, H. Van (1975) 'A field study of stress experienced by operators supervising a highly automated process', Productivity and Man, IFAC Workshop, publ. RKW Frankfurt.

Drury, C. G. and Baum, A. S. (1976) 'Manual process control, a case study and a challenge', *Appl. Eng.* **7**, No. 1, pp. 3–9.

Duyfjes, G., De Jong, P. J., and Verbruggen, H. B. (1977) 'Questionnaire on applications of modern control theory to computer control in the process industry. Results and comments', Paper R-8, IFAC/IFIP Conf. on Digital Computer Application to Process Control, pp. 833–845, North Holland, Amsterdam.

Edwards, E. and Lees, F. P. (1975) *The Human Operator in Process Control*, Taylor & Francis, London.

Erschler, J., Roubellat, F., and Vernhes, J. P. (1974) 'Automation of a hydroelectric power station using variable-structure control systems', *Automatica* **10**, pp. 31–36.

Fertik, H. A. (1975) 'Tuning controllers for noisy processes', *Trans. ISA* **13**, No. 2, pp. 172–181.

Fertik, H. A. (1977) 'Feedforward control of glass mould cooling', *Automatica* **13**, pp. 225–234.

Fjeld, M. (1978) 'Applications of modern control concepts on a kraft paper machine', *Automatica* **14**, pp. 107–117.

Friedewald, W. and Charwat, H. J. (1980) 'Design of graphical displays for CRT's in control rooms', *Process Automation*, January, pp. 7–13.

Funk, G. L. and Smith, D. E. (1977) 'Estimating economic incentives for computer control systems – an application approach', IEEE 24th Annual Petroleum and Chemical Industry Conference, 12–14 Inst. 1977, Dallas, Texas, U.S.A., Publ.: IEE, pp. 17–24.

Funk, G. L. and Smith, D. E. (1978) 'An applications approach to estimating economic incentives for computer control systems', Industry Applications Society, IEE–IAS 1978 Annual Meeting, Toronto, Canada, Publ.: IEE, New York, pp. 550–555.

Gabriel, E., Leonhard, E., and Nordby, C. (1980) 'Microprocessor control of the converter-fed induction motor', *Process Automation* 1, pp. 35–42.

Gawthrop, P. J. (1977) 'Some interpretations of the self-tuning controller', *Proc. IEE* 124, No. 10, pp. 889–894.

Gertler, J. and Sedlak, J. (1975) 'Survey Paper. Software for process control – a survey', *Automatica* 11, pp. 613–625.

Gifford, J. D. and Leigh, J. R. (1969) 'Design of a control system for a hot strip rolling mill', *Automatica* 5, pp. 433–447.

Gilbart, J. W. and Winston, G. C. (1974) 'Adaptive compensation for an optical tracking telescope', *Automatica* 10, pp. 125–131.

Glattfelder, A. H., Huguenin, F. and Schaufelberger, W. (1980) 'Microcomputer based self-tuning and self-selecting controllers', *Automatica* 16, pp. 1–8.

Handbook of Instruments and Instrumentation, Reference 85461 0642, Trade and Technical Press Ltd., Morden, Surrey, U.K.

Harland, G. E. (1973) 'Design of model reference adaptive control for an internal combustion engine', *Measurement and Control* 6, pp. 167–173.

Harrison, T. (ed.) (1980) *Distributed Computer Control Systems*, Pergamon Press, Oxford, U.K.

Hofmann, W. (1980), 'Process automation with decentralised control systems', *Process Automation* 21, No. 1, pp. 3–7.

Hughes, F. M. and Mallouppa, A. (1976) 'Frequency response methods for nuclear station boiler control', *Automatica* 12, pp. 201–210.

Hunter, R. P. (1978) *Automated Process Control Systems: Concepts and Hardware*, Prentice-Hall Inc., Englewood Cliffs, New Jersey.

Ichikawa, A. and Runa, E. P. (1979) 'Sensor and controller location problems for distributed parameter systems', *Automatica* 15, pp. 347–352.

The Instrument Manual, 5th Edition (1975), United Trade Press Ltd., London, U.K.

Jacobs, M. (1977) 'Comparisons of tuning methods for temperature control of a chemical reactor', Proc. ISA Chem. and Pet. Instrum. Symp., California, May 2nd–5th, published by Instruments Society of America.

Jacobs, O. L. R. and Saratchandran, P. (1980) 'Comparison of adaptive controllers', *Automatica* 16, pp. 89–97.

Jefferson, C. P. (1979) 'Feedforward control of blast furnace stoves', *Automatica* 15, pp. 149–159.

Joshi, S. and Kaufman, H. (1975) 'Digital adaptive controllers using second order models with transport lag', *Automatica* 11, pp. 129–139.

Junker, B. (1980) 'New methods of air-conditioning control – state of the art and development trends' (German), *Regelungstechnische Praxis* 22, No. 8, pp. 251–259.

Källström, C. G., Åström, K. J., Thorell, N. E., Erikson, J., and Sten, L. (1979) 'Adaptive autopilots for tankers', *Automatica* 15, pp. 241–254.

Keviczky, L., Hetthessy, J., Hilger, M., and Kolostori, J. (1978) 'Self-tuning adaptive control of cement raw material blending', *Automatica* 14, pp. 525–532.

Khandheria, J. and Shunta, J. P. (1979) 'Adaptive sampling increases sampling rate as process deviations increase', *Contr. Eng. Feb.*, pp. 33–35.

King, P. J. and Mamdani, E. H. (1977) 'The application of fuzzy control systems to industrial processes', *Automatica* 13, pp. 235–242.

King, R. P. (1974) 'On-line digital computer control of slurry-conditioning in mineral flotation', *Automatica* 10, pp. 5–14.

Kinzle, P. A. (1973) *Thermocouple Temperature Measurement*, Wiley-Interscience, London.

Koblitz, W., Pettersen, O., Clout, P. N., Kneis, W., and Wiesner, G. (1979) 'Industrial real-time Fortran', EURO IFIP 79, North-Holland Publishing Company, Amsterdam.

Kochhar, A. H. and Parnaby, J. (1977) 'Dynamic modelling and control of plastics extrusion processes', *Automatica* 13, pp. 177–183.

Landau, I. D. (1974) 'A survey of model reference adaptive techniques – theory and applications', *Automatica* 10, pp. 353–379.

Lau, C. C. and Leigh, J. R. (1979) 'The modelling of an industrial distributed parameter process', Proceedings of the IFAC Symposium on Identification and System Parameter Estimation, Darmstadt, West Germany, September.

Lee, D. H. and Han, K. W. (1980) 'Analysis of human operators' behaviour in high order dynamic systems', *IEE Trans.* SMC-10, No. 4, pp. 207–213.

Leigh, J. R. and Li, C. K. (1981) 'Adaptive control of ship steering', Proceedings of the IEE International Conference, 'Control and its Applications', Warwick, March, 1981.

Leigh, J. R. and Muvuti, S. A. P. (1981) 'A practical control algorithm derived through functional analysis', Proceedings of the IEE International Conference, 'Control and its Applications', Warwick, March, 1981.

Lopez, A. M. *et al.* (1967) 'Controller tuning relationships based on intergral performance criteria', *Instrumentation Technology* 14, No. 11, pp. 57–62.

Luenberger, D. G. (1966) 'Observers for multivariable systems', *IEE Trans. Automat. Contr.* AC-11, No. 2, pp. 190–197.

Luque, E. E., Tirado, J. F., and Moreno, L. (1979) 'A modular microprocessor-based control system', EURO IFIP 79, North-Holland Publishing Company, Amsterdam, pp. 263–271.

Mann, C. K., Vickers, T. J., and Gulick, W. M. (1974) *Instrumental Analysis*, Harper and Row, New York.

Martin, J. (Jr.), Corripio, A. B., and Smith, C. L. (1975) 'Controller tuning from simple process models', *Inst. Tech.* 22, No. 12, pp. 39–44.

Mehra, R. K. and Wells, C. H. (1971) 'Dynamic modelling and estimation of carbon in a basic oxygen furnace', Proc. 3rd IFAC/IFIP Conference on the Use of Digital Computers in Process Control, Helsinki.

Melsa, J. L. (1975) *Computer Programs for Computational Assistance in the Study of Linear Control Theory*, McGraw-Hill Book Company, New York.

Munro, N. (1972) 'Multivariable systems design using the inverse Nyquist array', *Computer Aided Design* 4, No. 5, pp. 222–227.

Otomo, T., Nakagawa, T., and Akaike, M. (1972) 'Statistical approach to computer control of cement rotary kilns', *Automatica* 8, pp. 35–48.

Perron, M. and Ramez, A. (1977) 'A survey of control strategies in chemical pulp plants', *Automatica* 13, pp. 383–388.

Porter, B. (1969) *Synthesis of Dynamical Systems*, Thomas Nelson and Sons Ltd., London.

Rijnsdorp, J. E. (1976) 'Man–machine communication in computerised chemical plants', Symposium Europ. Fed. Chem. Engrs., Firenze, Italy.

Rijnsdorp, J. E. and Seborg, D. E. (1976) 'A survey of experimental application of multivariable control to process control problems', Chem. Process Control, Proc. of Eng. Found. Conf., Pacific Grove, Calif., Jan. 18–23, 1976, pp. 112–123.

Rijnsdorp, J. E. (1979) 'State of the art and international development in the process computer field', *Elektrotechn. and Maschinenbau (EUM) (Austria)* 96, No. 6, pp. 251–255.

Rouse, W. B. (1975) 'Design of man–computer interfaces for on-line interactive systems', Proceedings of the IEEE, special issue on *Interactive Computer Systems* 63, No. 6, pp. 847–857.

Rouse, W. B. (1977), 'Human–computer interaction in multi-task situations', *IEEE Transactions on Systems, Man and Cybernetics* SMC-7, No. 5, pp. 384–392.

Sano, A. and Terao, M. (1970) 'Measurement optimisation in optimal process control', *Automatica* 6, pp. 705–714.

Schmidt, G. and Swik, R. (1974) 'Automatic hover control of an unmanned tethered rotor-platform', *Automatica* 10, pp. 393–403.

Sebakhy, O. A. and Wonham, W. M. (1976) 'A design procedure for multivariable regulators', *Automatica* 12, pp. 467–478.

Shilston, P. and Honess, J. (1974) 'Development of motorspeed control systems', *Electronics and Power, Drives and Motors* supplement, pp. 5–20.

Shinskey, F. G. (1977) 'The stability of control loops with and without decoupling', IFAC Conference on Multivariable Technological Systems, University of New Brunswick, July 5th–8th, pp. 221–230.

Shinskey, F. G. (1978) *Energy Conservation Through Control*, Academic Press, New York.

Singh, M. G., Drew, S. A. W., and Coales, J. F. (1975) 'Comparisons of practical hierarchical control methods for interconnected dynamical systems', *Automatica* 11, pp. 311–350.

Singh, M. G., Elloy, J. P., Mezencev, R., and Munro, N. (1980) *Applied Industrial Control*, Pergamon Press, Oxford.

Sira (1971) *Instruments in Working Environments*, Adam Hilger Ltd., London.

Sood, M. and Huddleston, M. T. (1977) 'Tuning PID controllers for random disturbances', *Instrum. Technol.* 24, No. 2, pp. 61–63.

Tustin, A. (1947) 'The nature of the operator's response in manual control and its implications for controller design', *J. IEEE* 94, Pt. IIA, pp. 190–207.

Tyssø, A., Bremo, J. Chr., and Lind, K. (1976) 'The design of a multivariable control system for a ship boiler', *Automatica* 12, pp. 211–224.

Tyssø, A. and Bremo, J. (1978) 'The design and operation of a multivariable ship boiler system', *Automatica* 14, pp. 213–221.

Wells, C. H. (1970) 'Optimum estimation of carbon and temperature in a basic oxygen-furnace', Preprints of the Joint Automatic Control Conference, Georgia Tech., Atlanta, June.

Wellstead, P. E., Prager, D. L., and Zanker, P. (1979) 'A pole assignment self-tuning regulator', *Proc. IEE* 126, No. 8, pp. 781–787.

Wellstead, P. E., Prager, D. L., Zanker, P., and Edmunds, J. M. (1979) 'Pole zero assignment regulators', *Int. J. Control* 30, pp. 1–26.

Wittenmark, B. and Aström, K. J. (1980) 'Simple self-tuning systems', International Symposium on Adaptive Systems, Bochum, West Germany.

Index